BUYING AND SETUP GUIDE TO HIGH - TECH ELECTRONICS

by
William Barden, Jr.

Radio Shack®

A DIVISION OF TANDY CORPORATION
FT. WORTH. TEXAS 76102

Editing: Janet Laurie
Artist: Mike Chiavetta
Graphics Design: Thyla Smith

First Edition
First Printing - 1988

Library of Congress Catalog Card Number: XX-XXXXX
ISBN 0-929449-00-2

Table of Contents

Preface

Do you want to assemble a super stereo system, but don't know how to go about it? Undecided about which television or monitor is best? Are you afraid of hooking up all those cables on a VCR? Would you like to install your own home security system, but think it will prove too complicated? Are you generally puzzled by connectors, plugs, jacks, cables, and wiring?

Buying and Setup Guide to High-Tech Electronics provides some of the answers to your questions about the many different types of electronic systems found in your local Radio Shack Store - what they are, which ones are best, and how to hook them up once you have them. Among the topics covered are:

- Audio systems: Making sense out of those specifications for stereo amplifiers, FM receivers, record turntables, cassette decks, and speakers. How to connect the system components. How to record your own high-quality material.

- Television, VCRs, and Camcorders: How to buy a good television or monitor and video cassette recorder and connect them to your television antenna or cable system. How to create your own television tapes using camcorders.

- Home security: What a home security system will do for you. How to easily install home security devices that will protect your home from burglary and fire.

- Telephone equipment: How to choose and install your own low-cost telephones, answering machines, and phone dialers, and save money in the process. The advantages of cellular telephones in your car.

- Car electronics: Everything you need for your car - CB radios, car stereos, and radar detectors. A guide to selection and installation.

- Electronic music keyboards, toys, and test equipment.

- A primer of home wiring - how to test your own home wiring for safety. Explanations of puzzling electrical and electronic terms.

- Battery buying guide.

There are truly some amazing electronic devices that will make your life easier, safer, and more enjoyable. *Buying and Setup Guide to High-Tech Electronics* will help you in the basics and guide you through the interconnections.

Chapter 1
Basic Components of a Stereo System

We all know *what* a stereo system does - it produces beautiful music, whether it's Bach or Rock. Sometimes, though, it's difficult to figure out what the different components of a stereo system are, which units to buy, and how to hook the parts together to get those beautiful noises.

A Basic System

There are several different types of stereo systems. Figure 1-1 shows the simplest type. It includes a stereo *receiver/amplifier*, an *antenna*, two *speakers*, and some interconnecting *cables*. With this system you can receive FM *stereo* or AM *monaural* broadcasts of any type (monaural is another word for "single channel", instead of the two channels you have with stereo).

The receiver/amplifier in Figure 1-2 includes both a radio receiver and an amplifier in one package. This unit is often just called a "receiver". The radio receiver receives FM or AM radio signals and converts them to *audio* - sounds that can be heard by humans. The signals produced by the receiver are very weak, similar to the level of sound you'd hear over a telephone handset. The amplifier in the receiver/amplifier unit amplifies these weak signals into levels strong enough to power large speakers.

A Separate Receiver and Amplifier

Sometimes the receiver and amplifier are two separate units. A receiver/amplifier may be just as good the two-unit system, saves space, and is more convenient to use. Some people prefer a separate receiver and amplifier, however, so they can "mix and match" different products. The receiver unit in this type of system is often called a *tuner*, and the amplifier is called an *amplifier*, rather than a receiver/amplifier.

Antenna

The antenna in Figure 1-1 may be built into the receiver. This is usually the case when the receiver is used for reception of AM broadcasts. However, for good reception of FM broadcasts, an *external antenna* is usually required. In strong signal areas, such as large cities, the external antenna may be as simple as the *folded dipole* antenna shown in Figure 1-3. In outlying areas (greater than about 40 or 50 miles, depending upon the terrain in your area), you'll probably need a more powerful antenna, such as the one shown in the figure.

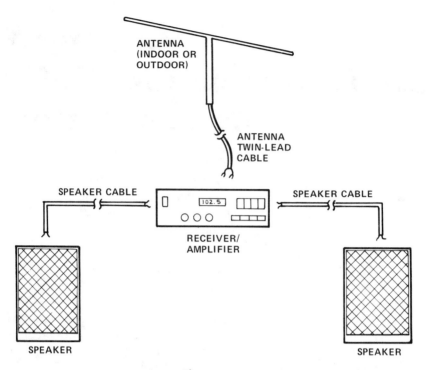

Figure 1-1.
A Simple Stereo System

Figure 1-2.
Receiver/Amplifier
Courtesy of Radio Shack
A Division of Tandy Corporation

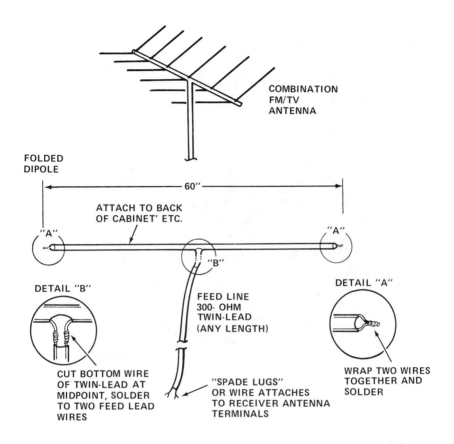

COMBINATION
FM/TV
ANTENNA

FOLDED
DIPOLE

|← 60" →|

ATTACH TO BACK
OF CABINET' ETC.

"A"   "A"

"B"

DETAIL "B"   DETAIL "A"

FEED LINE
300- OHM
TWIN-LEAD
(ANY LENGTH)

CUT BOTTOM WIRE
OF TWIN-LEAD AT
MIDPOINT, SOLDER
TO TWO FEED LEAD
WIRES

"SPADE LUGS"
OR WIRE ATTACHES
TO RECEIVER ANTENNA
TERMINALS

WRAP TWO WIRES
TOGETHER AND
SOLDER

Figure 1-3.
Antennas for AM/FM

Speakers

The speakers are sometimes built into less expensive stereo systems. However, for better sound quality, most stereo systems use separate speakers. For stereo, two speakers are needed. When stereo is recorded, the end product is two channels, one for each ear of the listener. The result is a "wrap-around" sound.

Any speaker will work, and produce much better sound than the table model radios of the 1950s. However, the speakers are one of the most critical parts of a stereo system. Generally, the larger and more expensive the speakers, the better the sound quality will be, particularly in the bass notes (see Figure 1-4).

Figure 1-4.
Stereo Speakers
Courtesy of Radio Shack
A Division of Tandy Corporation

Cables

In this simple system, only three *cables* (sets of wire) are needed, as shown in Figure 1-1. If you're using an external antenna, there's a cable that goes from the receiver to the antenna. There are also two two-conductor wires that run from the amplifier output to the speakers. We'll tell you how to hook them up later. For now, though, let's just say that they are inexpensive *speaker wire* that can be as much as dozens of feet long. The types of audio cables that are used in stereo systems are described in Chapter 4.

A Stereo System with a Turntable

The stereo system above allows you to play FM stereo broadcasts or AM monaural broadcasts off the air. The turntable shown in Figure 1-5 allows you to play records as well. Although records have lost some of their popularity due to tape cassettes, there are tens of thousands of different

Figure 1-5.
Turntable
Courtesy of Radio Shack
A Division of Tandy Corporation

records available, ranging from early 78 *rpm* (*revolutions per minute*) recordings to 33 1/3 and 45 rpm records that became popular in the 1950s and remained so to the current day.

Turntables are discussed in Chapter 2. For now, though, let's just say that a turntable can always be connected to the basic stereo system without buying additional equipment.

A Stereo System with a Cassette Tape Deck

Although some people prefer records, *audio cassette tapes* are extremely popular. Audio cassettes are of two basic types, *eight-track cartridges* and *cassette tapes*. The eight-track cartridges were once more popular than cassette tapes, but most equipment is now made for cassette tapes.

Although some *tape decks* come with a built-in amplifier, most are designed to be connected to a separate amplifier. A typical tape deck is shown in Figure 1-6. The tape deck allows tapes to be played back through an amplifier and speakers. The amplifier amplifies the small signal level of the cassette deck to listening volume.

Cassette tape decks also allow you to *record* programs in stereo from the receiver and other system components. Any station that can be heard on the receiver or any record that is played on the turntable can be recorded on cassette tape and then played back at a later time. The tape deck can also record sound from CD (compact disk) players, television, or video cassette recorders. The quality of recorded cassettes is very good, but depends mainly on the original material. Cassette tapes allow you to record up to 120 minutes of material on two sides.

Certain types of cassette decks allow you to copy one cassette onto another or to play two tapes in sequence. These cassette decks are called *dubbing decks* or *dual decks*.

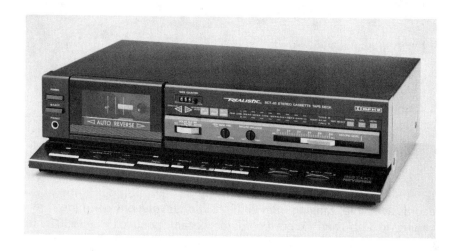

Figure 1-6.
Cassette Tape Deck
Courtesy of Radio Shack
A Division of Tandy Corporation

A Stereo System with a CD Player

Another addition to a stereo system is a *compact disk* player, as shown in Figure 1-7. *CDs* or *compact disks* are recorded *digitally* - signals are recorded by a series of tiny pits along the surface of the disk, which are then read by a laser beam. The presence or absence of a pit represents an "on" or "off" condition; a group of these on/offs are decoded into one single instant of sound for playback.

Figure 1-7.
Compact Disk Player
Courtesy of Radio Shack
A Division of Tandy Corporation

CDs are very popular - not so much as cassette tapes, but increasing in popularity. The advantage of compact disks is that there is no noise associated with the recording. Cassette tapes have some "hiss", and mistreated records have "pops" and scratches. However, CD recordings are characterized by noise-free reproduction, even at low volumes. CDs never deteriorate with reasonable use, as there are no parts that touch the CD surface during playback. CDs, however, are more expensive than cassettes or records. In addition, you cannot *record* on current CDs - only the CD manufacturer can do that.

CD players are connected to the amplifier of a stereo system with two cables. The amplifier does not need special converters to accept the CD player output.

Equalizers

Equalizers (Figure 1-8) are an optional piece of equipment for a stereo system that adjust the sound frequencies of your stereo system. The ideal sound system reproduces sound just as it was recorded. However, there are many factors that affect this ideal - your own hearing limitations, the amount of sound absorbing material in your listening area (drapes and soft, bulky furniture absorb sound, while hardwood floors reflect sound), the quality of your speakers, and so forth. Equalizers let you "custom-tailor" the sound to your own surroundings, equipment, and hearing. In some cases, the improvement is dramatic; in other cases, the equalizer does not add much to an existing system. (Almost all amplifiers have controls that will allow you to adjust the amount of *bass* and *treble* sound output.)

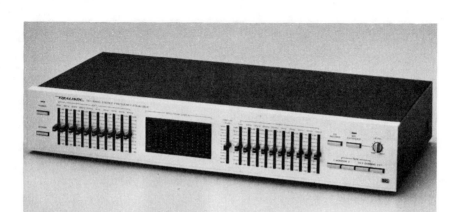

Figure 1-8.
Equalizer
Courtesy of Radio Shack
A Division of Tandy Corporation

A complete, full-blown stereo system for the audio fanatic is shown in Figure 1-9. It includes a receiver/ amplifier, FM antenna, turntable, cassette deck, compact disk player, equalizer, several speaker systems, and external connections to sound from television and VCR. All components of the system are available at Radio Shack.

Amplifier Buying Guide

There are certain *specifications* and buzzwords which are used to describe the performance of a stereo amplifier or receiver/amplifier. We'll describe them here.

Power

The first thing you will encounter upon reading amplifier specifications is a blurb such as "100 WATTS PER CHANNEL!". *Watts* are a measure of *power*. Just as a 100-watt light bulb is brighter than a 50-watt bulb, a 100 watt per channel amplifier is more powerful than a 50 watt per channel amplifier. The power here translates directly into sound *volume*.

How much power is required in an amplifier? Believe it or not, one or two watts of sound power produce a fairly loud audio signal. Chances are, you'll be using only 10 watts per channel or less in playing back even the loudest music. Then why are 100 watts needed? For one thing, the 100 watts are a *reserve* of power that is available for peaks in the sound. Drum crashes or loud brass call for more power without *distortion*, and the extra power of

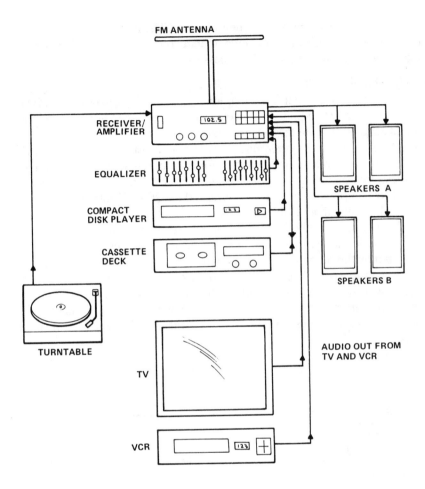

Figure 1-9.
Complete Stereo System

the amplifier provides it. It's like overdrive in an automobile. Usually you're cruising along at 55 or 65 miles per hour, but you may need a burst of passing power!

Power specs may be deceiving, however. Doubling the power - from 50 to 100 watts - *doesn't* double the sound volume. Doubling the power only increases the loudness by three decibels (3 dB), about three increments of volume.

The maximum power required for an average size room with typical furniture, is about 15 watts per channel! However, you may want to drive several sets of speakers, use the system in a large room, or maybe you just like your music LOUD. In that case, go for the extra power.

Extra power *is* important when playing CDs, as the CD signal has a much wider *dynamic range* of power than records or cassettes - from nearly inaudible audio to extremely loud passages.

Frequency Response

The next spec you'll see is *frequency response*. This spec is usually an easy one for most amplifiers to meet. The human ear can hear sounds from about 20 to 16,000 Hertz. The term *Hertz* is a unit that means the same thing as *cycles per second*, a measure of the frequency of a sound. An A above middle C note on a piano keyboard, for example, is 440 Hertz, while the highest note that a flute can play is about 9200 Hertz.

Most amplifiers can accurately reproduce 20 - 20,000 Hertz, plus or minus 1 dB, and that's sufficient for almost all music systems. The term + - 1 dB that you see after the frequency response refers to *decibels*, a measure of change in loudness. One decibel is a change in loudness that can *just* be detected by the human ear. If an amplifier has a response of 20-20,000 Hz + - 1 dB, and a complete range of frequencies is played through the system, you may *just* be able to tell the difference between the loudest and weakest response - if you know where to listen. When an amplifier has a good frequency response over a wide range, it is said to be *flat*, as shown in Figure 1-10.

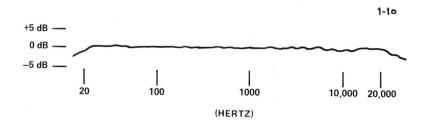

Figure 1-10.
Flat Frequency Response

Distortion

Power isn't everything. If an amplifier has 100 watts per channel, but distorts the music, you won't be able to stand the volume. *Harmonic distortion* measures the amount of distortion occurring when audio equipment adds *harmonic overtones* to the sound. If the harmonic distortion value is 0.5%, this means that 1/200 of the amplifier's output is harmonic distortion. Usually, if an amplifier specs state the harmonic distortion value it will be on the order of 0.2 to 0.3%, a not unreasonable figure. Lower-quality amplifiers probably won't provide the figure. Harmonic distortion should be less than 0.5% in a good amplifier.

Intermodulation distortion (*IM* distortion) is another type of distortion. It is caused by two tones mixing with each other. When music is played, many tones will produce this effect, causing a harsh and distorted sound. IM distortion should be less than 1% at all power levels.

Signal-to-Noise Ratio

Signal-to-noise ratio or *S/N* is a measure of how much noise there is in a system with no signal input. It's dependent upon the hum and noise you can hear with no signal input and the volume turned all the way up. Manufacturers would like to have systems without *any* noise, but there's always some. CDs have one of the best S/N ratios - about 90 dB. There's very little noise when a CD player has a completely quiet passage.

The lower the S/N figure, the worse the signal-to-noise ratio is. Most amplifiers have a S/N in the mid 80s, an acceptable value.

Equalization

When material is recorded on records, the frequency response is not flat. Bass is purposely deemphasized while treble is purposely strengthened, so that a record groove can be physically cut. The music or recorded material is recorded on a *curve* as shown in Figure 1-11. Likewise, when cassette tapes are recorded, a curve is used, since bass output from a recording head is weak compared to treble output. Amplifiers or cassette tape decks must have some means to change the frequency response back to

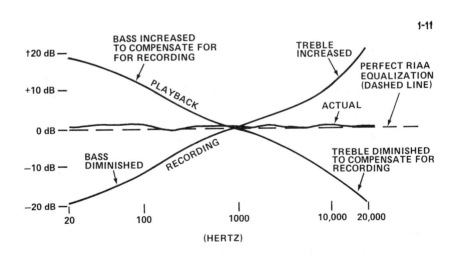

Figure 1-11.
Recording and Equalization Curves

a flat response. *Equalization* in an amplifier does this. Most amplifiers have automatic equalization built in for phonograph recordings. It's called *RIAA* (Recording Industry Association of America) equalization. Here again, amplifier specs are usually given as "RIAA + - 2 dB" or something similar. A + - 2 dB figure again indicates that the amplifier can match the RIAA curve fairly closely, compensate for the curve in recording, and reproduce the material with a flat response.

Channel Separation

When you play back material on one stereo channel, it is important that you can't hear it on the other channel. The *channel separation* figure shows how true this is. Once again, the figure is in dB. A figure in the high 40s is adequate, although the ideal would be something on the order of 90 dB. Usually this isn't of great importance, as music is playing through both channels at once, and you may not be able to notice whether there is a little *cross-channel interference*.

Receiver Buying Guide

The specifications for the receiver portion of a receiver/amplifier are identical to those of a separate receiver:

Sensitivity

Sensitivity is the measure of how well a receiver can capture weak stations. This is especially important for FM stations, which have a shorter range - usually only a little more than line-of-sight. Sensitivity is expressed in the signal level at the antenna terminals required for a given amplifier signal level. The signal is measured in *microvolts*, or millionths of a volt (radio signals are millions of times weaker than household wiring voltages!). Microvolts are abbreviated μv. The smaller the value, the more sensitive a receiver is. Most FM receivers are on the order of 1 microvolt sensitivity.

Signal to Noise

Signal-to-noise ratio for a receiver has the same meaning as for an amplifier. It's a measure of how quiet a receiver is. An FM receiver is not as quiet as an amplifier because of the nature of radio signals (there's inherent atmospheric noise that can't be helped). Typical S/N values for FM receivers are in the mid 70s. Again, the larger the value, the better the S/N.

Selectivity

Selectivity is the measure of how well a receiver separates one station from another. Since the FM band is divided into rigid frequency assignments, this is usually no problem. However, if you live near a strong station and are trying to receive a weaker one, you may hear the strong station even though you are tuned to the weaker signal. The selectivity signal is given in dB once again - the higher the value the greater the ability of the receiver to reject the unwanted signal. Typical selectivity values are in the mid 70s.

Other Receiver/Amplifier Features

Since the receiver/amplifier combination seems to be the most popular unit, let's consider some of the other features you may find in this integrated unit.

Digital Vs. Analog Tuning

One of the most noticeable differences between receiver/amplifiers is that some units are *digital* and some units are *analog*. Analog receivers are like older table model radios. Tuning is done by rotating a knob, with the frequency of the selected station displayed on a slide-rule dial with a pointer. This type of receiver is called an *analog receiver* because tuning is over a continuous range, just as some automobile speedometers register a car's speed on a dial.

Newer receivers, however, are digital units. The frequency readout is on an *LCD* or *liquid crystal display* (green), or possibly on an *LED* or *light emitting diode* (red) display. The display displays digits in much the same way that a digital alarm clock shows digits.

The digital design, however, goes much deeper than just the frequency display. The receiver tuning is done by a sophisticated quartz *frequency synthesizer* in these units. The chief advantage of this type of tuning is not that the receiver can receive stations more accurately, but that it can be *programmed* to receive certain stations at the touch of a button. Typical units of this type can store a dozen or so stations, all instantly selectable.

Loudness Circuitry

When music is played softly, the bass response and high notes become washed out. This is due to the *loudness* effect of human hearing - the ear has a different frequency response for different volume levels. A *loudness control* helps restore the bass and treble at lower volumes by boosting the bass and treble.

Typical Receiver/Amplifier Controls

A representative front panel of a receiver/amplifier is shown in Figure 1-12. This particular unit is a digital unit, which generally has more controls, many related to setting up the station preselects and tuning.

The *digital readout* provides the frequency of the selected station. A red signal indicator indicates when the received station is broadcasting in stereo. A few FM stations broadcast only in monaural. The five LEDs of the

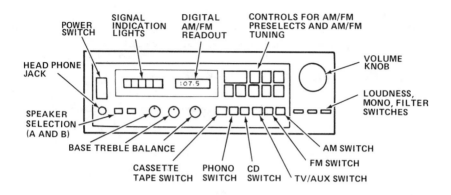

Figure 1-12.
Receiver/Amplifier Front Panel

signal strength indicator indicate the strength of the received signal. There are two sets of "up/down" tuning buttons, marked < and >. Automatic tuning allows you to search for the next station on the band. Manual tuning lets you control the tuning without the automatic search. Six buttons control the station *preselects*. Up to six frequencies can be stored in the memory of the unit in AM and up to six in FM by pressing a button followed by the MEMORY button. The station frequency is then stored in internal memory and kept there even when power is turned off (a small internal battery is used to retain the memory). Thereafter, just pressing the right button will switch to one of the stations.

A *VOLUME control* adjusts the volume. A *BALANCE control* lets you adjust the volume for the two channels, achieving the proper balance for the room, speaker placement, and so forth. *BASS* and *TREBLE controls* let you increase or decrease the amount of bass and treble to compensate for recordings or stations that are out of balance or to adjust for the room and surroundings. A *MONO* switch lets you make the two stereo channels common for recordings or stations that are not recorded in stereo. A *LOUD-NESS control* switches in the loudness equalization mentioned above.

An *MUTE* control turns off the speaker output when the receiver is tuned between stations. If this is not done, the resulting hiss and noise can be quite annoying, especially at 100 watts per channel!

The remaining controls are used to control the *source* of the sound. The *AM* or *FM* controls select either AM or FM radio. The *PHONO* button selects a turntable input attached to the system. *CD/AUX* and *TV/AUX* buttons select a CD player you may have attached to the system or output from a television or video cassette recorder. (These attachments are described in another chapter.) Note that all of these buttons are really *switches* that automatically select a different set of cable inputs attached to the rear of the unit. The *TAPE MONITOR* selects the cables for a cassette tape player you may have connected to the system without disabling any output so that you can simultaneously listen to and record FM, AM, CD, television, records, or other sources. The tape monitor is a way of rapidly checking the recorded material while it's being recorded.

Output of the receiver/amplifier goes to either speakers or *headphones*. Stereo headphones are a convenient way of listening to material in private during late evening hours or other occasions. Two switches - *SPEAKER A* and *SPEAKER B* - route the output of the amplifier to one or two sets of speakers. Usually the A speaker is one in the vicinity of the amplifier, while the B speaker may be at a remote location, say a pair of utility speakers in the back yard of your house.

Rear panel connections are described in Chapter 4.

Price

Typical receiver/amplifiers range in price from about $100 to over $1000. Very good quality units are to be found in the $400 area.

Chapter 2
Speakers, Headphones, Turntables, CD Players and Tape Decks

Once you have the basic stereo receiver/amplifier unit or units discussed in Chapter 1, you can add the additional sources of sound such as CD (Compact Disk) players and cassette tape decks described in this chapter. But first, you'll need one more essential component of a stereo system - speakers.

Speaker Basics

There is a wide range of speakers available for stereo systems in both price and size. In general, the larger the speaker, the better the bass sound quality. The more expensive the price, the better the sound quality as well. For stereo sound, two speakers are required, and they should be the same type for the best matching of the two channels. Suitable speakers range in price from about $100 to $600 per pair.

Speaker Enclosures

Speaker enclosures (cabinets) are not only an attempt to make the speakers look better. They are required. If a speaker is used without a cabinet, the sound wave from the rear tends to cancel the sound wave at the front of the speaker. Putting the speaker inside a cabinet eliminates this effect. However, once a speaker is put inside a cabinet there's another problem. Low frequencies (bass) need a large volume of air inside the cabinet to accurately reproduce the sounds. At the same time, there's a certain frequency in the bass range at which cabinet *resonance* occurs. Resonance produces a "booming" at certain bass frequencies. Therefore speaker systems need to be *tuned* by air ducts, channels, and the dimensions of the cabinet. Speaker enclosure design is a black art and there are dozens of different speaker designs!

Woofers and Tweeters

Within each speaker cabinet, there are often two speakers, a *woofer* and a *tweeter*. The woofer reproduces low frequencies, while the tweeter reproduces high frequencies. A *cross-over network* divides the frequency range sent by the amplifier into two groups of frequencies, one for the woofer and one for the tweeter.

In general, for best sound reproduction, a large woofer is needed for frequencies below 40 Hertz. Without such a woofer, the very lower frequencies will not be reproduced. That's why you'll see physically large cabinets in the best speaker systems, just as a tuba is a large instrument that produces low-pitched sounds. However, some speaker designs use smaller speakers with very good bass response.

On the high frequency end, the problem is simpler. Tweeters are physically small, just as flutes and piccolos are tiny and produce high-pitched sounds. Typical tweeters are three to four inches in diameter. Frequency response for inexpensive tweeters may go as high as 40,000 Hz.

Some speaker systems will have a *midrange* speaker as well, intermediate in size between the woofer and tweeter. In these speaker systems there are three ranges of sounds and an additional midrange *crossover point* for the crossover network.

The number of speakers inside an enclosure varies from one to ten or twelve. More speakers do not necessarily produce a better sound.

Speaker Power

In addition to speaker size, there's another important factor to consider when buying speaker systems - speaker *power*. When the term *speaker power* is used, it really refers to the ability of the speaker to handle the power generated by the amplifier. If the amplifier is a 100 watt per channel amplifier, the speaker must be able to handle at least 100 watts, about the same power as a 100 watt light bulb. Even though the average sound output may be four or five watts, *sound power peaks* (cymbal crashes and drums) may be much greater. If a speaker system cannot handle the power peaks, the sound may be distorted, or the speakers may even burn out.

Mini and bookshelf type speakers should not be used with heavy duty amplifiers that produce 50 or 100 watts of audio power, although they are fine for lower power amplifiers.

Speaker *transient response* refers to the ability of the speaker to handle short, loud noises, such as cymbals or drums. Listening to the same material over several speaker systems is a good way to test transient response. The sound should be sharp and distinct.

Speaker Impedance

Most speakers are eight ohms *impedance*. Impedance is a characteristic defining the electrical resistance at various frequencies. Most amplifiers will handle eight-ohm speakers without problems. In general, you should not use *less* than eight-ohm speakers connected to amplifiers that require eight-ohm speakers.

Speaker Sound Reproduction

After all the care taken to make amplifiers reproduce sounds exactly, with a flat frequency response, it's a shame to admit that the frequency response of most speakers is far from flat. A typical speaker frequency response curve is shown in Figure 2-1. The speaker will reproduce sound over all frequencies, but the sound volume is not *uniform* over the range.

Fortunately, the ear can't distinguish the somewhat rough frequency response of speaker systems that well. Because there are no absolutes as there are in amplifiers, however, one of the best ways to judge a speaker system is by listening. Choose the price range you can afford and then listen to all candidates. Even a $40 pair of speakers produces excellent sound quality compared to the sound systems of 20 years ago, and the quality is improving from year to year.

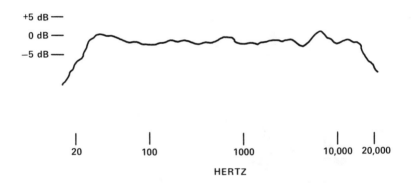

Figure 2-1.
Typical Speaker Frequency Response

Speaker Placement

Speakers should be placed so that the listening area is midway between the speakers. Good stereo requires that the two channels be at least six to eight feet apart, although this depends upon how close you are to the speakers. High frequencies are more directional than low and mid frequencies; it may help to angle the speakers so that they are pointing into the listening area.

Use the balance control on the amplifier to adjust the sound output from each channel. Due to sound adsorption and reflection by furniture, hard-surfaced floors, and drapes, the two channels may not be perfectly balanced without adjustment. The individual speakers may also have midrange and tweeter adjustment controls to match the speaker output to the room.

Connecting Speakers

Refer to Chapter 4 for tips on speaker connections.

Headphones

Headphones (Figure 2-2) can be used in place of speakers for privacy. Because the headphone speakers are right next to the ear, much less power is required for headphone sound output. Because of this, the headphone speakers can be much smaller and more responsive to sounds. Very good sound reproduction at any volume can be obtained from headphones in the $30 to $50 range. Another advantage of headphones is that the two channels of the headphones are almost perfectly matched - there is nothing to absorb or reflect the sound from the headphone speakers to your ears. Many headphones will outperform speakers costing ten times as much.

Figure 2-2.
Headphones
Courtesy of Radio Shack
A Division of Tandy Corporation

Make certain when buying headphones that they are stereo and have the proper sized plug for your amplifier. Most of the more expensive headphones designed for stereo systems use a 1/4-inch plug. Some of the less expensive variety designed for personal cassette players have a 1/8-inch plug; these should generally not be used for larger stereo systems, but can be adapted to the 1/4-inch size with a *plug adapter* if necessary.

Turntable Buying Guide

Turntables are used for most record types. However, newer turntables sometimes will not play the 78 rpm variety, so if your collection has these shellac pressings from the past, make certain you buy a turntable that has this capability. Turntables range in price from about $80 to $200.

Changers Versus Turntables

Many audio experts look down upon *record changers*. In a record changer a record drops onto another record while it is spinning. The grinding action creates a good chance for wear and introduction of noise. Turntables play one record at a time and even go so far as to provide soft *cueing* in which the tonearm is gently placed on the record surface.

Drives

Older turntables used a *belt drive*, a belt of rubber that went around the motor drive wheel and also around the wheel of the turntable axis. Belts, however, tend to wear and lose their tension, resulting in uneven speed. Newer turntables have *direct drive* motors, which drive the turntable directly, without an intervening belt. The direct drive type has a more constant speed and better sound quality as a result.

Rumble is a measure of motor noise in a turntable. The mechanical linkages of a motor are reproduced by the tonearm and appear as a rumbling noise. Like other specifications in audio systems, rumble is in dB - the higher the figure (disregard the minus sign), the less pronounced the rumbling noise. A figure of -65 dB or so is acceptable.

Wow occurs when the speed of the turntable varies at a slow rate and is noticeable by a change in pitch of a musical note. *Flutter* is also a variation in speed of the turntable, but occurs at a much faster rate - thousands of times per second. To detect wow or flutter, play a record containing a long sustained tone; there should be no noticeable variation in the tone - no warble - in a good turntable.

Tonearms

Recordings are essentially a grove in a vinyl material. This grove takes the form of a spiral. When *master recordings* are made, a movable arm creates the spiral groove by moving in towards the center of the master disk. The ideal turntable should have a *tonearm* that reproduces this movement exactly on playback. However, many turntable tonearms pivot around an axis and there is some *tracking error* as the arm follows the record groves, as shown in Figure 2-3.

More expensive turntables have a tonearm that duplicates the movement of the master recording turntable exactly, producing more precise sound. For casual listening, the difference between the two types of tonearms is not that great - you may not be able to detect any difference at all on most recordings. However, since the *linear* tonearm *does* eliminate tracking error and is not that much more expensive than the normal turntable, you may want to consider this type of turntable.

Cartridges

There are two basic types of turntable *cartridges*, the device that translates the wavy patterns of the record groove into low-level audio sound - *ceramic* cartridges and *magnetic* cartridges. Ceramic cartridges are found on less expensive turntables and require more weight on the record, causing record wear. Magnetic cartridges require less weight and provide more exact sound reproduction. The magnetic cartridges result in a lighter tonearm (a *low mass* tonearm), which tracks more easily and produces less record wear than ceramic cartridges. Magnetic cartridge signal levels are lower than crystal cartridges, however, and require *preamplification*. However, most newer amplifiers provide a magnetic phono input with built-in preamplifier.

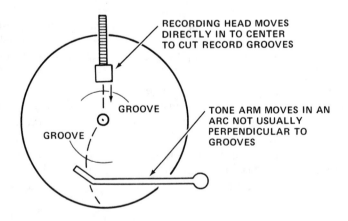

Figure 2-3.
Tracking Error in Turntables

Ceramic cartridges are found in less expensive, lower-quality turn-tables.

Connecting Turntables

Refer to Chapter 4 for details on connecting turntables.

CD Player Buying Guide

Compact Disk players (CDs) are a recent development. They offer noise-free, high-quality recording with disks that are read by a laser beam. From 60 to 75 minutes worth of material can be recorded. CD players are currently priced at about $150 to $500. The increased cost of most of the players is because of built-in *intelligence* that allows you to program the unit to play tracks in a certain sequence.

Frequency Response

The frequency response of CD players is excellent, thanks to the record-ing method used on the CDs. Sound is first converted to digital form - a series of ones and zeroes, or on/off conditions, as shown in Figure 2-4. The on/off conditions are then used to burn a pit or not burn a pit in the disk surface. The pit/no pit state is detected by a laser beam on the CD player, the digital data reconverted to audio form, and then fed to an amplifier. The quality of the reproduction is dependent upon how many bits - on/off conditions - are used in the CD player to reconstruct the data. Some less expensive CD players use 14 or fewer bits instead of the more common 16 bits. In CD players, "the more bits, the better".

Typical frequency response is 5 to 20,000 Hz and flat to within 0.5 dB.

On the con side: Some experts question whether the 44,100 samples per second of CD recording is enough to capture all the *overtones* of music. (Sampling rates must be at least twice the highest frequency to be recorded.)

Signal-to-Noise Ratio

The signal-to-noise ratio of CD players is also excellent. Because sound is represented by a discrete on/off condition, there is virtually no extraneous noise, except for the noise produced by the circuitry that reconstructs the audio signal, which is minimal. Typical S/N (signal-to- noise) values are 90 dB - as good or better than many amplifiers.

Dynamic Range

Dynamic range is a measure of the volume range a CD player can cover - how much greater the loudest sound is compared to the most quiet sound. This figure is dependent upon the number of bits used in reconstructing the signal and is therefore a good check on that circuitry. The dynamic range should be 90 dB or greater for better quality units.

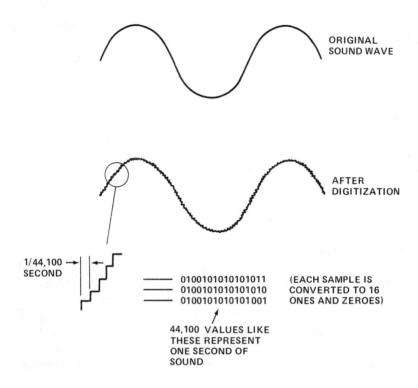

ORIGINAL
SOUND WAVE

AFTER
DIGITIZATION

1/44,100 →| |←
SECOND

0100101010101011   (EACH SAMPLE IS
0100101010101010   CONVERTED TO 16
0100101010101001   ONES AND ZEROES)

44,100 VALUES LIKE
THESE REPRESENT
ONE SECOND OF
SOUND

Figure 2-4.
CD Recording Method

Other Audio Specifications

CD players, like turntables, have wow (slow variations in motor speed) and flutter (faster variations in motor speed), but in almost immeasurable amounts. Channel separation between the two stereo channels is also excellent - typically 90 dB or more, due to the digital nature of the recording.

Intelligent Features

Most CD players, like VCRs, have good to excellent electronic circuitry for reproducing the audio sound. The main differences between CD players is in the amount of *intelligence* they provide. CD disks themselves have much more "data" built-in. Whereas a record has only sound data along a spiral track, the CD disks have information that identifies a track number. This means that any track can be selected automatically and that tracks may be "programmed" to play in sequence - the CD player can find any track with its search circuitry.

Also included with many CD players is a *wireless remote control*, often called a *wireless remote*. This is a remote control similar to many found on televisions - an infrared beam that has control information in it. To change to the next selection, for example, just aim the beam towards the CD player and press a button. The wireless remote is a handy feature to have, but adds $30 or so to the price of the CD player.

Some CD players are actually CD *changers*, allowing five to ten CD disks to be held in a magazine. All of the disks may be played through in sequence, or selections may be programmed - tracks from various disks can be played in a predetermined sequence. CD changers cost quite a bit more than CD players, however - from $200 to $500 more. If playing 10 hours of uninterrupted music is important to you, though, the higher price may be worth it.

Connecting CD Players

See Chapter 4 for details on connecting CD players.

Cassette Tape Deck Buying Guide

Unlike turntables or CD players, cassette tape offers the ability to *record* music and sound in addition to playing it back. There is quite a wide variation in cassette tape material - ferric tape, chrome tape, and others. Add to that a noise reduction scheme called *Dolby* and you have a lot of factors to consider when buying a cassette tape unit. Cassette tape decks are priced from about $90 to $400.

Basics of Cassette Recording

Cassette recording is a development of magnetic recording in general. Magnetic recording techniques first used *wire*, then *reel-to-reel* tape, then 8-track *cartridges*, and now *cassette*. Reel-to-reel tape recording is still in use for expensive studio equipment. In this type of system, 900 to 3600 feet of tape is held on a 7-inch (and larger) reel and travels through a playback/record head to a second reel.

Cassette tapes are small versions of a reel-to-reel scheme. Sixty to 120 minutes worth of tape is held on a single cassette. Each cassette records in stereo on two sets of tracks, as shown in Figure 2-5. In less expensive cassette players the tapes must be manually flipped over to playback the second track. In higher-quality recorders, there is an *auto-reverse* feature which changes the direction of the tape and positions the *tape head* to the second track on the tape.

Cassette tape is vinyl plastic coated with a magnetic material. The original material was ferric oxide, commonly known as rust! However, newer tape is coated with a coating of chromium dioxide ($CrO2$) or metal. These coatings produce higher-quality sound at the expense of more *head wear*.

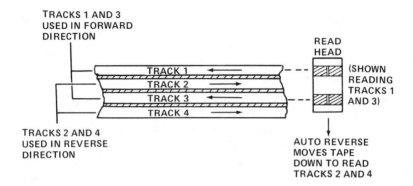

Figure 2-5.
Cassette Tape Playback

Most new cassette decks can playback and record on any type of tape - ferric oxide, chrome, or metal. Selections between the tape types are made by switches.

Dolby Noise Reduction

Tape recording is characterized by a background *tape hiss*. The hiss is very noticeable during quiet periods in the tape. However, several years ago, Dolby Laboratories developed a *noise reduction* scheme that reduced tape hiss considerably, and extended the dynamic range of recording. The Dolby scheme requires special electronic circuitry to playback tapes recorded *under the Dolby system*. All cassette decks except the least expensive incorporate Dolby noise reduction. Dolby "A" is used for professional recording, while Dolby "B" and "C" are used for consumer cassettes.

Frequency Response

Frequency response of cassette tapes depends upon the material the tape is made of in addition to the quality of the recording electronics. Here are some general guidelines:

- Standard ferric oxide tape: 40 - 14,000 Hz

- Chromium oxide tape: 40 - 15,000 Hz

- Metal tape: 40 - 16,000 Hz

High-frequency response can be as much as 2,000 Hz less in lower-quality recorders. Tape recording response is not as flat as most amplifiers. The frequency figures are + - 3 dB, which translates to lower output on the high frequencies.

Signal to Noise

The *signal-to-noise ratio* (*S/N*) is on the order of 60 to 70 dB, with 70 dB being best for most cassette recorders *under Dolby*.

Other Cassette Specifications

Just as in turntables, there is *wow* and *flutter* in cassette decks. Wow occurs when the tape speed changes at a slow rate and flutter occurs when tape speed changes at a faster rate, for example, vibration past the tape head. Typical wow and flutter specifications are 0.1%, with a lower figure being better. Distortion is also expressed as a percentage, with a lower value being better. Typical harmonic distortion is 1%.

Types of Tape Decks

There are several different types of cassette tape decks:

• *Cassette playback decks* allow you to play back prerecorded tapes, but not to record. Output goes to an amplifier.

• *Cassette decks* allow you to play or record one tape, with the audio output going to an amplifier.

• *Dual-cassette decks* or *dubbing decks* allow you to copy one tape to another tape or to play two tapes in sequence with output to an amplifier.

Dual cassette decks (dubbing decks) are generally not as good as a single tape deck of equivalent cost. However, dubbing decks are very handy for copying cassettes, and most will do the copying in about half the normal playback time. Generally there is some loss in quality with each dubbing of a tape - a third or fourth generation tape is not as clean as the first generation original.

Some dubbing decks have *auto-reverse* to copy both sides of a tape automatically and some will play two tapes in sequence.

Typical Cassette Deck Controls

Figure 2-6 shows the front panel of a representative tape deck. There are two microphone jacks, one for each channel of recording. (Recording techniques are covered in Chapter 3.) There is also a headphone jack. Although the audio output of cassette decks is not powerful enough to drive speakers (it will drive small speakers at very low volume), it is more than adequate for headphones. The headphones can be used to monitor the recording.

Two *recording level* knobs control the input signal level. The recording level knobs are used in conjunction with two *peak level meters*. Some

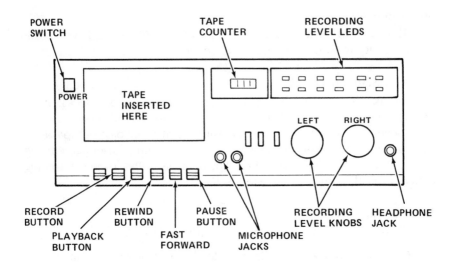

Figure 2-6.
Typical Cassette Tape Deck Front Panel

cassette decks use actual meters for the recording level, but five LEDs (light emitting diodes) are also used to indicate when the input signal level from both channels is at the proper level - too little and the tape will be noisy - too much and the signal will *overload* the input electronics.

A *RECORD* button starts the recording process. A *PAUSE* button will temporarily stop the process at a finger press. Tape is indexed by a *tape counter*. This is a digital counter similar to an automobile odometer. Using the *REWIND* button will rewind the cassette tape either to the start of the tape or to the 0000 setting of the tape counter. The tape counter can therefore be used as a way to mark a spot on the tape to come back to by resetting it to 0000. A *FAST FORWARD* button starts a high-speed tape movement forward (as opposed to a rewind) that helps to locate any point on a cassette.

Different *biases* or electrical characteristics are used for the three tape types - ferric, CrO2, and metal. Two buttons enable selection of the proper biasing for the tape type being used. If the proper bias is not used, tape playback will sound flat or unnatural.

A Dolby button selects the Dolby noise reduction electronics for tape playback. Tape playback is selected by a *PLAYBACK* button. The *RECORD* button is *interlocked* so that it cannot be easily pressed by mistake, destroying all or a portion of a cassette.

Rear Panel Connections

See Chapter 4 for information on connecting cassette cables.

Chapter 3
Recording Techniques

Recording in stereo is easy with a cassette deck. It's as·simple as plugging in two microphones and pressing the *RECORD* button. However, there's a lot that can be done to make your recordings more professional, both in techniques and equipment. We'll discuss both topics in this chapter.

Recording Tape

Cassette tape comes in several types and several lengths. *Standard ferric* tape is the original tape type and is also the cheapest. It is fine for general recording, providing a balanced, uniform sound, especially at low frequencies. High frequencies, however, will be diminished. However, the recording cannot be better than the *source material*, so at times, standard ferric tape will be as good as the most expensive tape. For example, if you are recording from *AM* radio, the frequency response is of much lower quality than an FM source, and the ferric tape will be as good as the original.

The second tape type is *chromium dioxide (CrO2)* tape. This is a higher quality tape than ferric tape and is about twice as expensive. What you'll get for the extra money is extended high frequency performance and also improved signal-to-noise ratio - softer sounds can be heard over tape hiss. Another type of tape in the same category is *chrome equivalent* tape. This type of tape contains no chrome, but is less expensive than the CrO2 tape, while still retaining better high-frequency response and signal-to-noise ratio.

Still more expensive is the third tape type, *metal tapes*. Metal tapes are very good in the high frequencies. They also increase the dynamic range of recording, allowing very soft to very loud sounds to be recorded. Sound reproduction in the lower frequencies is similar to ferric tape, however.

The best way to decide on the tape type you'll want to use is to actually try all of the tape types with the same recorded material. Record two or three minutes from the same record or another cassette, and then compare. Much of the decision rests on your own personal preference.

Tape lengths range from C30 (15 minutes per side, 30 minutes total) to C120 (60 minutes per side, 120 minutes total). Although it would seem like the best bargain is the C120, the tape used here is thinner and may stretch. Another consideration is the *length* of the recorded material. Although you can continue recording in the other direction automatically, there will be a noticeable break as the cassette deck changes direction. If you are recording from records or CD, this is no problem, but you may have to time off-the-air material accordingly.

There are bargain cassettes available at low prices. However, some of these are factory rejects that couldn't pass the specs. Bargain cassettes may have *dropouts* due to thin or missing magnetic coating, or other problems. It's best to stay with a reputable manufacturer who has good quality control.

Recording From Records, Off the Air, or Other Sources

Remember that the recording can never be any better than the original source material. When recording, make certain that the source signal is as strong and clean as possible. If recording from records, this means the records should be physically clean. There are *record cleaning cloths* and other cleaning schemes that remove dust from record grooves and prevent pops and clicks. An amplifier with a *pop* filter also will reduce record noise.

When recording off the air, consider beefing up your antenna system to get the strongest signal possible. The antenna signal can never be too strong, unless you happen to be next door to the transmitter!

Two of the best sources of recorded material are compact disks and video cassette tapes. Compact disks are especially clean and make excellent recordings. Sound from stereo video cassettes requires a stereo VCR, of course, but is usually also excellent. Although the law is still somewhat clouded in the area of taping original material, it goes without saying that recordings of this type must be for your own personal use and cannot be sold as pirate copies!

When recording, remember that what you're hearing out of the loudspeakers as you're playing back the source material while recording does not affect the signal level of recording. The signal to the recorder from the amplifier *TAPE OUT* jacks is a constant, low-level signal that is *not* controlled by the volume control on the amplifier. The controls that determine the signal level into the recorder are the *RECORDING LEVEL* controls for left and right channels on the recorder. Adjust these controls so that the needle on a recording level meter is below the red zone. It may peak into the red zone occasionally, however. If you have LED indicator lights, adjust the controls so that the 0 DB LED is on frequently; occasional peaks into higher levels are all right.

Make certain that the two channels are balanced and peak at about the same levels. If you find that the recording level is too high or low and want to adjust the levels up or down, do it smoothly over a minute or so, preferably during loud passages, where it won't be noticeable to the listener when it's played back.

Chapter 4 describes the cabling for recording from various sources.

Recording On the Spot

Two microphones are needed for stereo recording of music or other material. These microphones should be well matched - they should be purchased together. There are many different types of microphones, some of them not suitable for recording.

A *carbon microphone* is good for only low-quality sound. It is commonly used in telephones and two-way radios.

A *crystal microphone* is an inexpensive mike with good frequency response, but is somewhat sensitive to overloading.

A *ribbon microphone* is a sensitive microphone with good sound quality. High frequency response is not as good as some other types, however. Ribbon mikes are not as rugged as other types and may be damaged by loud sounds.

A *dynamic microphone* is a widely used microphone for public address and recording with a fairly level frequency response. It is less subject to overload from loudness and is rugged.

An *electret microphone* or *condenser microphone* is a very sensitive microphone that provides a level frequency response and is ideally suited for high-quality recording. A small battery is required to power an internal preamplifier, as the signal level output is low.

A *lavalier microphone* is designed to be used as a tie-clip mike for lecturers. Because a user's chin cuts off many of the high frequencies, this type of microphone has a high-frequency *boost* that accentuates high-frequency response.

Microphones have different *pickup patterns* - they respond to sound from different directions. A *shotgun* microphone, for example, is used to concentrate the pickup into a tight beam and eliminate sounds from anywhere outside of this beam. The shotgun mike is used for news conferences where it's aimed at the speaker.

An *omnidirectional* pickup pattern picks up sound equally from all directions, as shown in Figure 3-1. A *bidirectional* pickup pattern picks up sound from the microphone front and rear. A *unidirectional* or *cardioid* (heart-shaped) pickup pattern picks up sound from the microphone front only.

There are other features to be found on mikes. A *pop filter* or *windscreen* eliminates pops in speech (such as the "pah" in pronouncing the word "pop") and wind noises. *Shock mounts* eliminate vibration transmitted through the mike stand.

Recording Environment

Avoid rooms with a lot of reflectivity of sound when recording. Rooms with hardwood floors and many windows will reflect the sound back and create an echo effect. The ideal conditions would be a *studio* environment. Recording studios use these tricks to control sound:

- Walls are not at right angles to each other

- Walls are well insulated, so that external noises - sirens, barking dogs, etc. - cannot leak in

- Sound absorbant materials such as drapes, heavy carpeting, and large cloth-covered furniture are used to reduce reflected sounds.

- Portable *baffles* are used to reduce sound reflections, or, in some cases, to increase sound reflections.

Speakers shouldn't be used while recording, as they will result in *feedback* - high-pitched squeals as the amplified sound is feed back into the microphone. Instead, headphones should be worn to monitor sound output. That's why you see recording artists wearing headphones while doing "takes". It's the only way they can really hear what is going on.

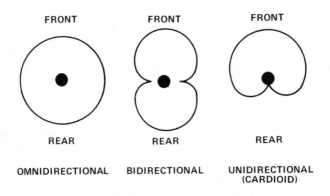

Figure 3-1.
Microphone Pickup Patterns

Recording Voice

To record a single performer in stereo, two cardioid mikes placed at angles about nine inches apart can be used, as shown in Figure 3-2. A *dual-head* mike provides this pattern in a single unit. Dynamic mikes will give a good response. The performer should be six to 18 inches away from the mikes, and the mikes should have a windscreen or pop filter to eliminate breath pops. Place the mikes above or below the mouth of the performer at a 45 degree angle.

When several performers are involved, they can be positioned on either side of the mikes, but still "in the pattern".

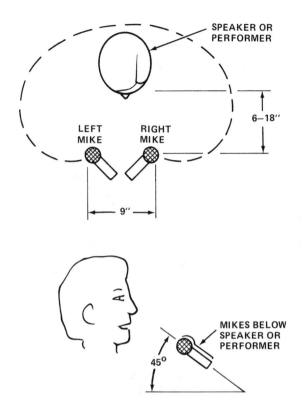

Figure 3-2.
Recording a Single Performer

Recording Small Musical Groups

The highest-quality microphones should be used in stereo music recording, to obtain the best high-frequency response. Even though an instrument may not have the range of notes into extreme high-frequencies, the *overtones* of the instrument will. Overtones (harmonics) give the instrument its color and distinctiveness and are present at high frequencies.

Cardioid microphones placed about four to ten feet apart and aimed towards the center of the group will give good results, as shown in Figure 3-3. The distance from the microphones to the front of the group should be about four to six feet. Omnidirectional microphones can be used as well, provided that there isn't a great deal of sound reflection from behind the mikes (use sound absorbant material placed behind the mike if this is the case).

Two mikes can be used for piano, one a foot above the sounding board, and the other pointed at the raised lid.

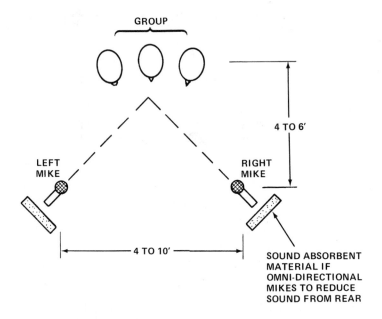

Figure 3-3.
Recording Small Groups

Mikes for guitars should be placed one to three feet from the guitar hole, or angled toward the low or high strings as desired. Mikes for banjos should be one to two feet from the center of the banjo head.

Violins and violas should be recorded with a mike a few feet away, with the mike aimed at the center of the instrument.

Brass instruments such as trumpets are loud in comparison to others. Record them from one to five feet away.

Woodwinds should be recorded with mikes a foot or so from the finger holes of the instrument.

One or several singers should have mikes placed a few inches to a few feet away, depending upon the room surroundings.

Recording Large Groups

For large groups, such as choirs or orchestras, two high-quality microphones should be used again, but spaced further apart - about eight to 15 feet and placed well back from the group - 15 feet or more, as shown in Figure 3-4. Some experimentation is necessary here, and test recordings are a good idea. Two omnidirectional mikes may not cover the entire choir or orchestra. One solution is to use two mikes together for each channel - four mikes in all - and combine the channels with a *mixer* (described below). Another solution is to use omnidirectional mikes and eliminate any reflected or audience noise from the back or sides by sound-absorbant material placed around the mikes.

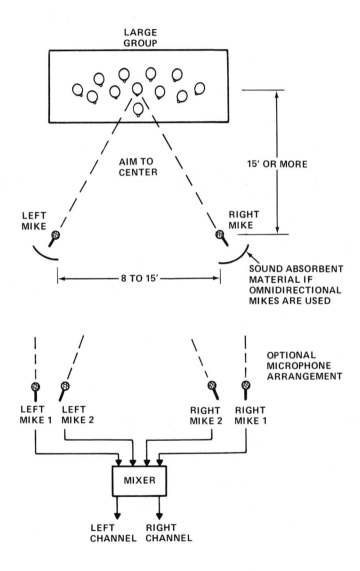

Figure 3-4.
Recording Large Groups

Mixers

Mixing equipment does not refer to Oster blenders in recording, but its the same idea. Mixers are devices (see Figure 3-5) that allow you to combine several audio channels together into a single channel or several channels.

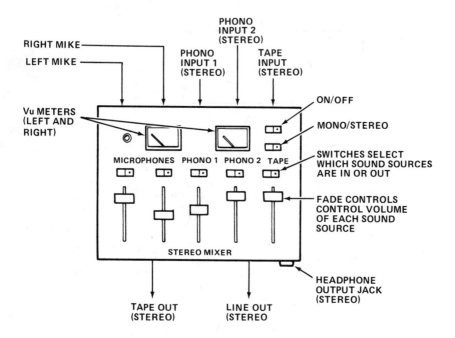

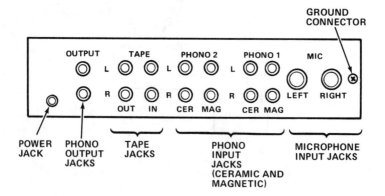

**Figure 3-5.
Typical Mixer**

When a recording is made of an orchestral group, movies, or television, there may be many channels of audio that are recorded. These are mixed by audio engineers to provide the proper blend of instruments, performers, and sound effects. You can do the same with inexpensive audio equipment.

Each mixer has a series of input channels and output channels. Several input channels can be combined and the audio level of each one can be adjusted by a mixing slide control, which changes the level of each input from zero to maximum.

The mixer may incorporate *VU* (*volume unit*) meters to obtain the proper signal levels, or may have LED level indicators similar to those found on cassette recorders. VU meters measure average sound energy in much the same way that a human ear does.

In addition to mixing channels, some *sound processors* may act as equalizers so that you can adjust the bass or treble (or frequency ranges in between) to a proper balance to correct recordings that are too "boomy" or "tinny".

Chapter 4
Hooking up Audio
Equipment

In this chapter we'll talk about how to hook up audio equipment such as receivers, cassette recorders, and the audio portion of video cassette recorders. Most of the connections are easy, but we can offer such hints as the proper wire to use, cable length, and descriptions of plugs and jacks.

Wire Gauge

The size of wire used in audio connectors is measured in a system called *AWG* or *American Wire Gauge*. In this gauge, *smaller* diameter wires have *larger* AWG values and *larger* diameter wire has *smaller* AWG values. House wiring wire, for example, is typically 14 or 16 gauge, but wires to batteries used in small electronic devices are typically 22 gauge. Use Figure 4-1 below to gauge the size of wires if in doubt.

A rule of thumb about wire sizes: It's all right to use larger diameter wire, but never use smaller diameter wire than what is called for. Smaller diameter wire will produce more heat or may degrade the signal.

Cabling Safety

In all modern audio equipment, there is no problem in handling *audio cables*. Audio cables do not have high voltages present and are safe to disconnect and connect into various pieces of equipment. Always keep audio cables separate from *power cables* that plug into house wiring, however. Usually this is no problem, as most audio equipment has a single *line cord* that plugs into a wall outlet. It's a good practice to turn off all audio equipment when connecting or disconnecting cables, however, to avoid possible damage to circuits caused by high-powered "pops" or *transients* caused by connecting or disconnecting cables.

Speaker Wire

Low-loss speaker wire is 18-gauge stranded copper wire. A smaller diameter gauge may be used for short runs. The *loss* term refers to power lost due to the resistance of the wire. This power is burned off in the form of minute amounts of heat. About 1/160 of the power output of an amplifier will be lost in a *ten* foot run of 18-gauge copper speaker wire. About 1/16 of

GAUGE	BARE WIRE WITHOUT INSULATION (ACTUAL SIZE)	DIAMETER (INCHES)
0		.3249
5		.1819
10		.1019
14		.0641
16		.0508
18		.0403
20		.0320
22		.0253
24		.0201
26		.0159

Figure 4-1.
Wire Gauges

the power output of an amplifier will be lost in a *100* foot run of 18-gauge wire. When *100* feet of *24*-gauge copper wire is used, about 30% of the power output will be lost in heat. You can see that for short runs, the diameter of the wire is not all that important, but for longer runs it becomes more of a factor.

Stranded copper wire should be used to make the speaker wire more flexible. If *solid* copper wire is used, the wire may break with repeated bending.

There are two wires for every speaker, usually connected by *quick connect* tabs. The two wires for each speaker are connected to the two *RIGHT* or *LEFT* speaker connections on the amplifier, as shown in Figure 4-2. Speaker wire can be purchased that has *tinned pins* on each end. The ends of the wire have been soldered so the strands will not separate. The ends of this type of wire can be easily inserted into the spring-loaded tabs of the speakers.

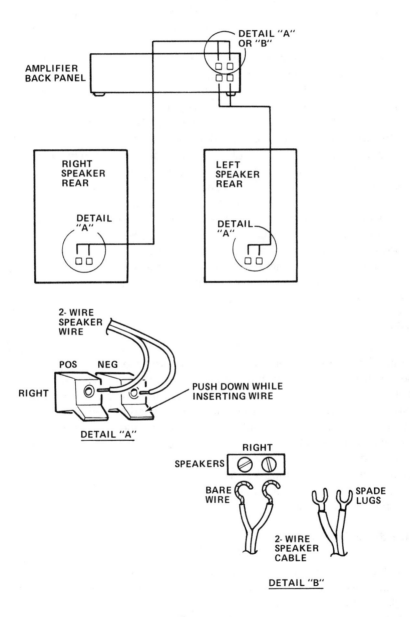

Figure 4-2.
Connecting Speakers

To connect bulk wire, separate the two wires, cut off about 3/4 inch of the insulation from each wire, and stick into the spring-loaded tabs on the speaker. To connect to the amplifier side, stick into the speaker tabs in the same way. Some amplifiers will have screw-type terminals. In this case, form the wire into a U-shape in a clockwise direction, and screw down the wire, making certain that no individual strands cross between the two screws.

One important point: Speakers are *phased* so that the speaker cones for both channels move in the same direction (in or out) with the same signal. If speakers are *out* of phase, there is no electrical harm, but the sound will not be as bright and will seem diminished as one speaker's sound output will tend to cancel the sound output of the other speaker. Speaker connections usually are color coded with a red and black terminal; match the color code on both sets of speakers, use a color-coded speaker wire, and make certain that the amplifier connections of both sets of speaker wires match. If in doubt, try taking off one set of wires on one speaker and reversing the leads. You may notice an improvement in the sound output.

Speaker wire can be routed under floors and through ducting without too many problems. Just be certain that the wire does not come close to raw ac power (outlets or wires) and that the speaker wire is not exposed to the bare copper along the length. Speaker wire can also be coiled, but not too tightly. (If coiled on a roll it might act like an electrical *induction coil* and affect the frequency response of the system).

L-Pads

L-Pads are *volume controls* for speakers, installed between the output of the amplifier and speaker. An L-pad can control the volume of a remote speaker from no sound to full volume. (However, the amplifier volume control still is the "master control" that determines the maximum possible volume for the speaker output.) L-pads are designed to handle high-power signals, unlike the volume controls on the front panel of amplifiers. They also match the speaker *impedance*, or electrical load. L-pads look like volume controls, but are mounted on standard wall plates which may be installed in electrical wallboxes. A typical installation for an L-pad is shown in Figure 4-3. Use an L-pad with a high enough power rating for the amplifier and speaker combination in use.

Standard Audio Cables

Standard audio cables for connection of cassette decks, CD players, turntables, VCR sound output, television output, and other audio applications look like Figure 4-4. The cable is a shielded *coaxial* type with a center conductor surrounded by a woven metal shield. It's important to use a shielded cable, as the shield eliminates hum and noise from the low-level audio signals.

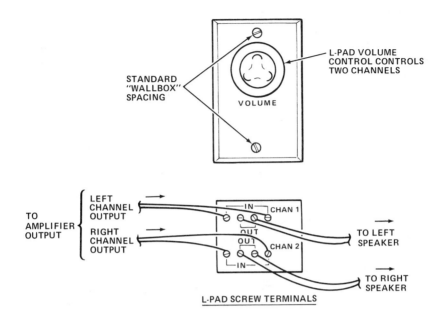

Figure 4-3.
L-Pad Installation

(THIS END NORMALLY
HAS ANOTHER PHONO
PLUG)

WOVEN METAL
SHIELD

OUTSIDE
INSULATING
JACKET

PLASTIC INSULATING
MATERIAL

CENTER CONDUCTOR
(SOLID)

RCA OR PHONO PLUG
(TYPICAL— MAY BE
ANOTHER TYPE OF
PLUG)

Figure 4-4.
Standard Audio Cable

These cables are often referred to as *audio patch cords* and are available at Radio Shack and other suppliers.

The connectors on the ends of the cables are known as *RCA* type connectors or *phono plugs*. You'll need cables with a phono plug on one end and a phono plug on the other end. (Some cables have a phono plug on one end and a phono *jack*, or female plug on the other end.) These cables are available in lengths from 3 to 12 feet, and you should have no problem in finding them the proper length for your installation. Use the shortest lengths possible to reduce system noise and to eliminate cable "rat's nests".

If you need to extend a cable, use a cable with a phono plug on one end and a phono *jack* on the other end. This cable can be inserted between two normal plug-to-plug cables.

The cabling for a typical full-blown stereo installation with receiver/ amplifier, turntable, CD player, cassette deck, television input, and VCR input is shown in Figure 4-5, together with cable types.

Receiver Antenna Connection

If you are using an external antenna, there are two possibilities for cabling to the receiver. The first type of antenna uses 300-ohm *twin-lead*. This type of cable usually has *spade lugs*, which can be connected to the 300-ohm terminals in the back of the receiver, as shown in Figure 4-6.

Another alternative is a more elaborate antenna that uses coaxial cable. In this case, the plug from the cable can be connected to a coax antenna jack, if one is available on the back of the receiver. If only a 300-ohm twin-lead terminal is available, you'll need a small device called a *matching transformer*. This device converts the 72-ohm *impedance* of the coaxial cable to a 300-ohm impedance twin-lead that can be connected to the terminals of the receiver as shown in Figure 4-7. Without the matching transformer, you will not get the best reception from the antenna.

AM/FM receivers generally have two sets of screw terminals for an antenna, one for an FM antenna, and one for an AM antenna. The AM antenna is usually not necessary, unless you're in a remote area. However, if you do need an AM antenna, the antenna is much less critical than an FM antenna. Often, just a piece of *hookup wire* (22- or 24-gauge stranded copper wire) of three to ten foot length may be connected to one of the AM screw terminals and will provide good reception of AM stations.

Headphone Cables

Headphones usually come with about ten feet of coiled cord. This may not be long enough to reach from an amplifier to your easy chair. A *phone plug to in-line phone jack* cable will let you extend the headphone cord to reach an additional ten feet or so. (You can add additional extensions as required.) Make certain that the phone plug and jack on the cable fit the size of plug on your headphones - most are the 1/4 inch type.

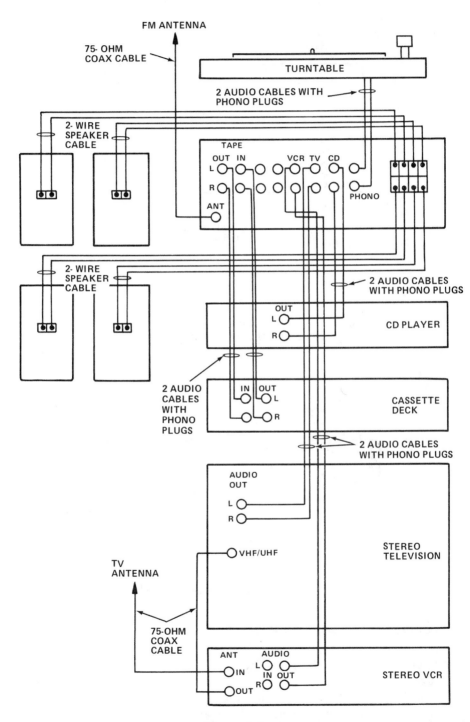

Figure 4-5.
Cabling for a Stereo System

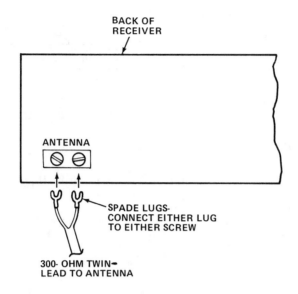

Figure 4-6.
Receiver Antenna Connections

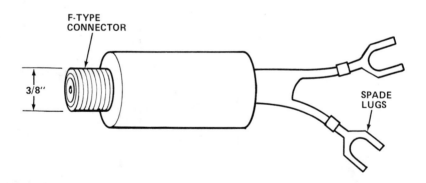

Figure 4-7.
Matching Transformer

Receiver/Amplifier Rear-Panel Connections

A representative rear panel for a receiver/amplifier is shown in Figure 4-8. Although at first the many jacks may seem confusing, they're all straightforward.

Jacks for stereo are labelled either *RIGHT* or *LEFT* or *R* and *L*. That's right or left as you face the speakers. Jacks are also labelled *IN* or *OUT*. "IN" means that a signal is coming into the amplifier from a turntable, cassette player, CD, television, or another audio source. OUT means that a signal is going out to a recording device, which in almost all systems is a cassette player.

The in and out audio signals are all at very low voltages, and there is no chance of receiving electrical shock from them. Also, it is generally not possible to "blow something out" by hooking up the wrong audio cables. The exceptions are the speakers, which may be damaged by strong signal levels.

Speaker connections are generally of the type in which a spring-loaded tab is pressed so that a stripped wire may be inserted into a hole. The tab is then released, and the wire is gripped by friction.

Antennas are connected to the system by either a *coaxial* type connector or *spade lug* connectors as described above.

Turntable Cabling

Two cables come out of stereo turntables, one for each channel of sound. They are connected to the *MAG PHONO IN* or *PHONO IN* jacks on the stereo amplifier. The *RIGHT* and *LEFT* cables of the turntable should be connected properly to the *RIGHT* and *LEFT* jacks on the amplifier.

4-8

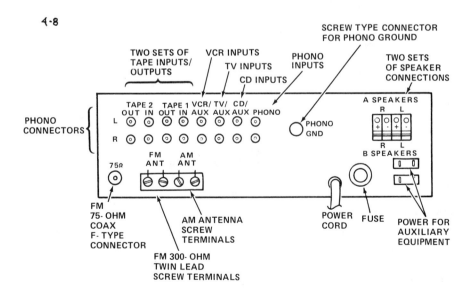

Figure 4-8.
Typical Receiver/Amplifier Rear Panel

CD Player Cabling

There are two audio cables that come out of a CD player, one for the left channel and one for the right channel. They connect to the *AUX* or *CD* inputs of an amplifier. The *RIGHT* and *LEFT* cables of the CD player should be connected properly to the *RIGHT* and *LEFT* jacks on the amplifier.

Cassette Deck Cabling

A representative rear panel for a cassette deck is shown in Figure 4-9. There are four jacks labeled either *LEFT* and *RIGHT* or *L and R. TAPE OUT* jacks provide the output signal to be connected to an amplifier *TAPE 1* or *TAPE 2* input jacks. *TAPE IN* jacks are used for the amplifier's *TAPE OUT* connections. All jacks can be interconnected between the cassette deck and amplifier using standard audio cables. TAPE IN and TAPE OUT are sometimes called LINE OUT and LINE IN.

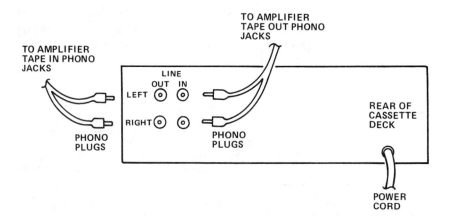

Figure 4-9.
Cassette Deck Cabling

Connecting Television and Video Cassette Recorder Audio

Newer television *monitors* (see Chapter 5) come with audio output jacks. If the television is stereo, there will be two *AUDIO OUT* jacks, one for each channel; monaural sets have one *AUDIO OUT* jack. The jacks are standard phono-plug type jacks that will fit audio cables. To record television audio, just plug the one or two audio cables into the television jacks and route them to the *AUX* input jacks of the amplifier and from there to the cassette recorder, as shown in Figure 4-10. Sometimes the amplifier will have a special *TV* input, and in this case, plug the cables into those one or

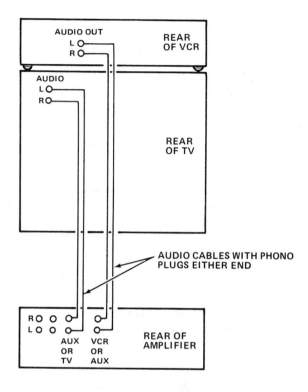

Figure 4-10.
Television and VCR Audio Cabling

two jacks. (If you're recording monaural television, you can switch to *MONO* on your cassette recorder to record only one channel.)

Video cassette recorders are similar to television monitors. They have one audio jack if they are monaural, or two jacks if they are stereo. The jacks are labelled *AUDIO OUT*. Route one or two cables from the VCR to the amplifier *AUX* or *VCR* inputs; from there, the normal cassette recorder output will capture the television sound when the amplifier is switched to the AUX or VCR input.

Microphone Cabling

Most microphones use either a 1/4-inch or 1/8-inch headphone plug, as shown in Figure 4-11. The 1/4-inch plug will fit mixer jacks or public address system jacks. The 1/8-inch plug may have to be converted to a 1/4-inch plug, however. An inexpensive 1/8-inch to 1/4-inch adapter will make the conversion. Since the audio inputs on amplifiers are generally phono jacks, you may also have to convert from the 1/8-inch or 1/4-inch headphone plug to a

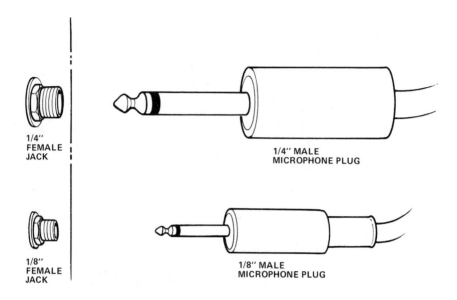

1/4"
FEMALE
JACK

1/4" MALE
MICROPHONE PLUG

1/8"
FEMALE
JACK

1/8" MALE
MICROPHONE PLUG

Figure 4-11.
Microphone Plugs and Jacks

phono plug; adapters are also readily available for this conversion. There's no electrical problem in making the conversion. Standard audio cable may be used to extend the "run" of microphone cables.

Getting Rid of Noise in Your System

There shouldn't be a great deal of hum or extraneous noise in your stereo system. If there is try these tips:

- Equipment with three-prong plugs is automatically grounded to house wiring ground. Other equipment may have a grounding connector, usually a wing nut connected to the rear panel. It often helps to reduce noise by connecting to the house wiring ground as shown in Figure 4-12.

- Also, test your wall outlets with a *circuit tester*, a small plug with indicator lights. It will tell you if your plug is properly grounded. Grounding not only reduces noise, but is a good safety precaution for using household appliances and electronic equipment.

- Test all cables. Does a hum appear or disappear when certain cables are jiggled? If so, try another cable - the suspect cable may have a broken shield connection, introducing hum into the audio line.

- Are your amplifier or receiver controls "scratchy"? Scratchy controls cause a static-like noise when rotated. If so, use television tuner cleaner on the controls. This is a chemical that comes in a spray can. Spraying in the chemical and then rotating the controls for a few minutes will often cure the problem.

- Is your recorded sound distorted? If so, you may be recording at too high a level. Once recorded, there's no way you can retrieve the audio, but make certain the *RECORD LEVEL* meter or led lights do not indicate recording in the red zone. *Audio levels* should not cause the meter or LEDs to be continuously in the red zone (above 0 dB).

Making Your Own Cables

It is possible to make your own audio cables. Radio Shack carries a complete line of plugs, jacks, and cables. *Shielded audio cable* should be used for all audio applications. *Solderless phono plugs* can be used on the

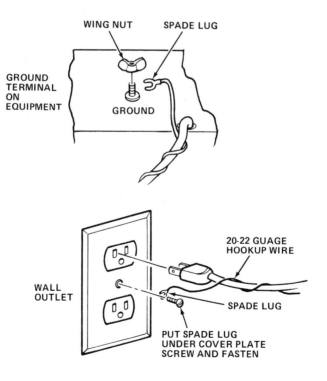

Figure 4-12.
Grounding Equipment

ends of the cable, or, if you're handy with tools, soldering is not that difficult. You'll need a *wire stripper/cutter* for the solderless type of connections and an additional *soldering iron or gun* and *rosin-core solder* for soldered connections (use only rosin-core solder to prevent corrosion).

The advantage to making your own cables is that they may be custom tailored in length. In addition, they are generally less expensive. When properly done, the cables you make will perform just as well as the ready-made variety.

Chapter 5
Television and Video Equipment

Television sets have come a long way since the first commercial sets in the 1940s - color and stereo sound have improved the picture and audio. Today's sets, however, offer many new features which make television viewing even more enjoyable. We'll look at many of these features in this chapter. We'll also describe monitors, large screen tvs, and pocket tvs.

Standard Televisions

The standard television signal is broadcast over the air from local stations. Since television frequencies are in the *VHF (very-high-frequency)* and *UHF (ultra-high-frequency)* range, reception is generally *line-of-sight* - if the transmitter is more than 60 or 70 miles distant, the signal quality deteriorates. At 120 miles it is very weak or non-existent.

The standard television picture has 525 lines on the screen. If you look closely at a television picture, you can see the actual screen lines. In fact, some of these lines are invisible, as they go off the edges of the picture. This is called *overscan* and is purposely done, to avoid a border around the picture on the screen. However, too much overscan is not good, as a portion of the picture is lost.

When a picture is transmitted, every other line of the 525 lines are transmitted in the first 1/60th of a second, and the other half of the lines are transmitted in the second 1/60th of a second. Because of persistence of vision, the lines *interlace* together, like intertwined fingers, to make up the entire picture (see Figure 5-1). The picture quality on television can never be any better than these 525 lines, at least with current television transmission (a new high-resolution television transmission is in the planning stage, but will be some time in coming, if at all).

The sound portion of television is *FM*, the same type of FM signal used for FM radio, which produces high-quality sound. Many stations are now broadcasting in stereo, which adds a new dimension to tv viewing.

Cable Television

Cable television now is used in over one-half the households across the nation. Cable television receives signals from satellites for such services as

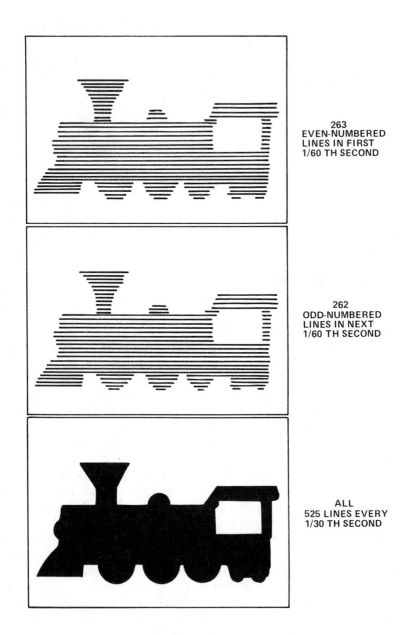

263
EVEN-NUMBERED
LINES IN FIRST
1/60 TH SECOND

262
ODD-NUMBERED
LINES IN NEXT
1/60 TH SECOND

ALL
525 LINES EVERY
1/30 TH SECOND

Figure 5-1.
Television Picture Interlace

Home Box Office, ESPN, and Super Station WTBS. It also receives signals of local television stations, or those in a neighboring city. All of these signals are then merged into one set of signals that are fed through a single cable and routed to many homes in an area (see Figure 5-2). Cable fees range from $10

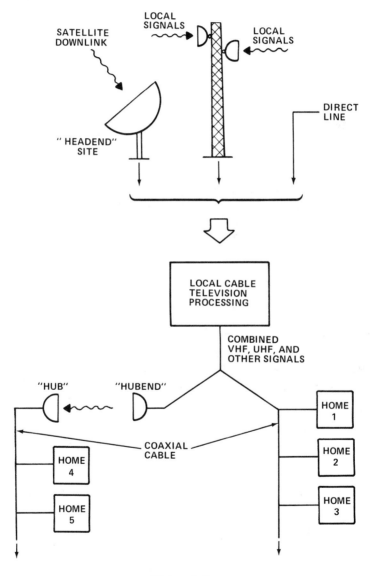

Figure 5-2.
Typical Cable Tv Operation

a month upwards, depending upon how many services to which you sub-scribe.

Cable television offers a wide selection of viewing - everything from music videos, to Congressional sessions, to local stations showing network television shows. Cable picture and sound quality is very good overall. However, off-the-air pictures and sound can be as good or better as cable with the proper antenna system.

Newer television sets now not only include the normal television channels, but also make provision for receiving cable television as well. This is called being *cable-ready*.

Satellite Television

There's a third type of television transmission as well, *TVRO (Television Receive Only)* television. With TVRO, a large satellite antenna and low-noise amplifier are used to receive signals from orbiting satellites that carry network television programming. These broadcasts are then received on an ordinary television receiver.

TVRO is an expensive ($900 - $3000) alternative to receiving only local television. However, picture quality is generally excellent and the system is perfect for isolated areas. However, many satellite services are converting to *scrambled* transmissions to prevent TVRO viewing unless you have a special decoding box and pay a monthly fee.

Television Vs. Monitors

There are two types of television systems in home use today. Television *receivers* are not too much different from the television sets of *The Honeymooners* days. They receive off-the-air signals and convert them into tv picture and sound. Television *monitors*, however, are used to view pre-recorded television played back on *video cassette recorders* and *laser video disk players*, in addition to being used with home computers and video games.

Pure *monitors* cannot receive television signals off the air or from cable. However, most consumer sets are *receiver/monitors* which combine both a standard television set and a monitor in one package. The advantage of a monitor is that it bypasses the *front-end* electronics of a normal television, which degrades signal quality. The resulting picture is much sharper and clearer than a television picture, which has been converted from an off-the-air signal.

Television Buying Guide

Here are some of the specifications and buzzwords used to describe televisions and monitors, together with guidelines for acceptable quality.

Tuners

Early television tuners used a manually rotated dial to select the channel to be viewed. New televisions use a *PLL (phased locked loop)* type of circuitry and a *quartz tuner*, which doesn't necessarily mean that the channel can be tuned in more exactly, but does make life somewhat simpler when selecting channels. Once this circuitry is in place, it enables channels to be selected by the digits of the channel number from a remote tuner in your easy chair! It also allows *programming* of channels - selection of often used channels stored in the memory of the television.

A *cable-ready* tuner means that no special converter box is required for connecting cable television. The converter boxes change the frequencies used on the cable to corresponding channel frequencies in your set, in addition to some decoding of special services. Cable-ready sets are a kind of "built-in" converter box. Stations on *subchannels* are automatically converted to the proper UHF or VHF channel. Typical cable-ready sets have 110 to 155 channel capability (some of these channels are cable subchannels).

In some cases cable services charge extra for the converter boxes, so a cable-ready set can save you money in that regard. Also, not having to go through another piece of equipment is an advantage, and it makes hooking up a television, video cassette recorder, and audio equipment a great deal easier.

Automatic Frequency and Color Circuits

Better sets have an *automatic frequency circuit*, in addition to *automatic color circuits* to tune in the channel frequency and adjust the color to standard values without manual tuning. These circuits can be overridden so that the channel can be fine-tuned manually and so that colors can be adjusted to the viewer's preferences.

MTS Stereo Sound

Some sets have built-in stereo sound capability. When a signal is received with stereo sound, the set automatically switches to stereo audio and sends the sound to two audio amplifier channels within the set. The amplified sound is then sent to the set's two stereo speakers. In addition to self-contained stereo sound (which is usually not that good because of the set's small dimensions and speaker placement), the audio output is routed to *AUDIO OUT* jacks on the set. The two channels of audio (right and left) can then be connected to a stereo amplifier and speaker system (see Chapter 4). The resulting sound quality can be very good. For this reason, many people make an *entertainment center* combining television, VCR, and stereo system.

If you are not connecting your television's audio output to a stereo, make certain that the audio output power of the set is adequate. Audio output power is measured in watts, the more the better. Two watts of audio power may not be enough for a large room and some sets are weak in this area.

SAP

SAP stands for "Second Audio Program". It is an "invisible" channel of audio superimposed on the regular television signal. This channel is used for second language broadcasting and other purposes. It can be decoded if the set has SAP capability, and replaces the normal audio. A newscast may be broadcast in both Spanish and English, for example. Selecting SAP switches to the Spanish language subcarrier. Not all stations use this subcarrier, of course, but in most major metropolitan areas there is a second audio on some of the channels.

Picture Quality

Picture quality for television receivers, like audio systems, can be judged best by viewing several sets with the same material. One good way to do this is to playback a video cassette recorder through several sets in sequence. The VCR will provide a stable picture. Playing back a VCR in this fashion is a good check on the *video electronics* portion of the television, but not of the *front-end electronics* that decode the picture from the off-the-air signal. Tune in the same station on several sets to do that.

One important piece of electronic circuitry is a *comb filter*. A comb filter is an involved electronics circuit that matches the color brightness (luminance), and color hue (chroma). On televisions without a comb filter, *ringing* occurs during intense colors of different hues and brightness. The effect of this is a series of noticeable zebra stripes across the picture. A comb filter greatly reduces this effect. Picture resolution is also increased with a comb filter.

The measure of how fine a detail a television can display is provided by the *horizontal resolution*, expressed in number of lines. As we mentioned earlier, a television picture is composed of '525 lines. The ideal television displays all of these lines sharply. However, most sets display a picture made up of only several hundred lines - each of these lines represent a merger of several of the original 525 lines. The more of these lines that can be individually displayed, the better the quality of the picture. A good set can display on the order of 300 lines, while an excellent set can display 450 lines or more.

A related measurement is the *video frequency response*, expressed in megaHertz (MHz). The higher this figure, the better. Each megaHertz compares to about 80 screen lines, so a set with a video frequency response of 5.0 megaHertz should be able to display about 400 screen lines.

Another factor affecting picture quality is how well the screen lines interlace together. If the lines do not mesh perfectly (remember, they fit together in two halves like interlaced fingers), you will see white areas between lines of the picture.

The *transient* response of a television defines how well a television picture displays outlines of objects. The worse the transient response, the fuzzier the edges will be on outlines of objects.

The *luminance* of a television is a measure of picture brightness. The greater the luminance, the brighter the picture will be. A luminance of 70 to 100 footlamberts is good for most sets, but projection televisions should have as high a luminance as possible because of the larger picture area. (Did you ever notice how bright small-screen televisions sometimes appear? - it's due to the condensed picture size.)

Black-level retention is another factor of picture quality. If a set has good black-level retention, black areas on the picture will appear a contrasty black and will not appear washed-out. Another factor here is good screen reflectance - the screen will not absorb room light and wash out the picture.

Adjacent channel interference may sometimes cause herringbone interference patterns in a picture. This may be true especially in the case of cable television, where adjacent channels (say channels 4 and 5) are of nearly equal strength. A set with good adjacent channel rejection will show no herringbone or ghost pattern on the screen.

Digital Signal Processing

Most all current televisions are *analog* sets, meaning that the signals are processed continuously by internal circuitry. *Digital* sets, however, break up the incoming television signal into small units of digital information, in a similar approach to the way a compact disk player reconstructs audio by discrete "ones and zeroes" of information.

Once the picture is in digital form, a great many tricks can be played with it. For example, the normal 525 lines on a television tube can be expanded to double or three times that number by reproducing each line twice or three times, There is no new picture information, but the overall effect is to "smooth out" the picture and make it appear to be of much better quality without visible lines. This technique is just being introduced, and is not available on many sets at present.

Another thing that can be done with the digitized picture is to make a *picture-in-picture* (PIP). In some more expensive sets, for example, you can view one channel on the full screen and simultaneously watch a smaller inset picture at a quarter or sixteenth of normal size. The inset picture may come from another channel or video cassette recorder (see Figure 5-3). Another option provided with this type of set is the ability to "freeze" the action of the inserted picture.

Also in the immediate future is the ability of digital sets to eliminate picture "ghosts", although no sets are currently available with that capability.

Figure 5-3.
Television Picture-in-Picture
Courtesy of NEC

Remote Control Smart Televisions

Even without digital signal processing, however, many television sets are much smarter than they used to be. Many sets now have a wireless remote control. Earlier sets used a remote control connected by a cable to the set to enable a viewer in an easy chair to switch channels, turn the volume up and down, and turn the set on and off. Today, virtually all televisions are "wireless" - the remote control uses a beam of infrared light to send a signal to the set to accomplish remote control functions, without a wire to trip over.

Direct-access remote control allows you to select any channel by number, rather than by stepping from channel to channel. Some sets allow programming your twelve most popular channels for *scanning* - you can step through these channels in sequence.

Many more expensive sets now have on screen text to help in tuning, channel programming, parental control (blocking certain channels from viewing by children), and other functions. In some cases, you can even label the channel with your own text. You could, for example, call channel two "CBS" or channel 18 "STKS" for "stocks".

Teletext is another option that is becoming available on television. Teletext provides information sent at the same time as normal television programming - stocks, weather, entertainment, travel, telephone numbers, and so forth (see Figure 5-4). Although not often available in this country, Teletext is widely used in some European countries. The information appears on the screen as a series of "pages". You can select any page you desire by entering a page number - since the pages are continuously broadcast, the proper page will appear on the screen in a matter of seconds.

Of course, to receive Teletext, it has to be available in your area, and you must have a set that is designed to receive it, or a special converter box.

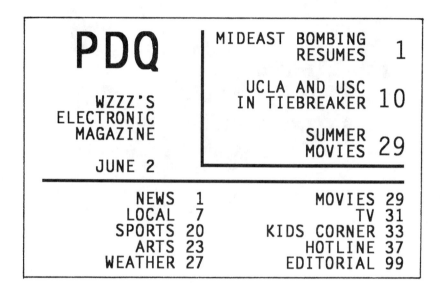

Figure 5-4.
Teletext

Large Screen Televisions

There are two basic types of large screen televisions, those sets with *large picture tubes*, and *projection systems*. Televisions with 25-inch picture tubes were once about the largest screens that you could buy. However, manufacturers now provide tube-type sets with tubes from 27 to 35 inches. Picture tube measurement, by the way, is *diagonal* measurement, across the tube from corner to corner.

Projection televisions provide pictures up to six feet in diagonal measurement (see Figure 5-5). Rear-screen projection tvs provide pictures up to about five feet diagonally. Rear-screen projection televisions are a single-piece unit. Two-piece units are less popular than they used to be, and project an image from a central television about the size of a small coffee table onto a separate screen. High luminance (brightness of picture) is

Figure 5-5.
Projection Tv
Courtesy NEC Electronics

important in all large-screen televisions. Get one with the greatest luminance level possible. Projection television prices range from about $1200 to
$3500 or more.

Miniature Televisions

On the other end of the scale are miniature televisions, sometimes
called *pocket tvs*. Some of these sets use a small conventional picture tube
that has been bent for small dimensions. Other sets use a *liquid crystal
display* (LCD) that can provide a small, flat screen. Sets are available both in
black and white and in color. Picture size ranges from about 1 1/2 inches to 3
1/2 inches (see Figure 5-6).

LCD display sets sometimes offer a poor viewing angle. If your viewing
angle isn't just right, the picture will disappear. LCD sets also do not offer the
highest resolution - the picture may appear grainy. Typical line resolutions
are 200 lines, about one-half that of the usual larger television. On the other
hand, LCD sets are usually viewable in direct sunlight - the light helps rather

than hinders the picture brightness. Tiny sets with conventional tubes may appear washed out in bright sunlight.

Reception is usually good, but may be subject to your orientation as you move around. Moving your body slightly, for example, may adversely affect reception on weaker stations.

Audio quality on pocket sets is usually not that good. Many sets have a built-in speaker, but because of the small dimensions of the set, the speaker will not give a good bass response. Almost all of these sets have a built-in headphone jack for private listening.

Pocket televisions are powered by batteries or by ac or dc adapters. The dc adapter may allow the set to be plugged into a cigarette lighter in a car, for example. Battery life is an important factor. Typical battery-powered pocket televisions use four AAA or AA batteries with battery life of about two to four hours. That's two to three dollars per viewing session, unless rechargeable NICAD batteries are used!

Television Antennas

All television sets depend upon a good antenna system for best picture and sound reproduction. If you have a cable television service, about all you'll have to do is plug in the cable to the television, and then tune to VHF and the lower-numbered UHF channels. (Cable service switches channels around somewhat so that the top ultra-high frequencies do not have to be

Figure 5-6.
Pocket Television
Courtesy of Radio Shack
A Division of Tandy Corporation

carried on the cable.). The cable service is likely to provide the local channels, channels from other cities, such as Chicago and New York, special services, such as C-SPAN, the Congressional Channel, and ESPN, the Sports Network, and subscriber movie channels, such as Home Box Office and The Movie Channel. The latter services are available for additional fees and in some cases, installation of a *decoding box* to unscramble the subscriber channels. Without the decoding box, you'll see a jumbled picture on the screen.

Cable connection details are described in Chapter 7.

If your set is "cable ready" you will be able to receive all channels from the cable and may not require a connecting box. If your set is not cable ready, you may require a connecting box even though you are not buying subscriber services. There is usually a switch on the television receiver so that you can switch to a CATV position, rather than from an off-the-air signal.

If you do not have cable, you may be able to use the antennas included with the set if you live fairly close to the center of a major metropolitan area (within about 30 miles). These are usually small wire antennas that clip to the back of the set (see Figure 5-7).

The further out you are from the transmitter and the more terrain that comes between the transmitting antenna and your set, the more you will need some type of receiving antenna. There are two basic types of antennas, for the two ranges of channels received on your television. VHF (Very High Frequency) refers to one set of channel frequencies and UHF (Ultra High Frequency) refers to the second set of channel frequencies. VHF channels are channels 2 through 13, while UHF channels are channels 14 through 83. VHF antennas are physically larger than the UHF antennas. Generally, the more expensive and larger the antenna, the better will be the resulting picture and sound. Most antennas combine a VHF and UHF antenna, eliminating the need for two separate antennas. Antenna setups are described in detail in Chapter 7.

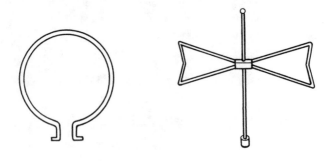

Figure 5-7.
Local Tv Antennas

Chapter 6
Video Cassette
Recorders and
Camcorders

Video cassette recorders (VCRs) have become increasingly popular - over 40 million were sold in the past three years in the United States alone. VCRs allow you to record from television or other sources and to play back your own recordings or commercially produced movies. We'll look at VCR features in this chapter. We'll also look at a hot new item - *camcorders*, combinations of video cameras and recorders. Camcorders are an alternative to home movie equipment and offer many advantages over that earlier way of capturing images of Bill Junior or Aunt Sadie.

VCR Formats

A few years ago, there were two formats used in video cassette recorders - *Beta* and *VHS*. Beta was promoted by Sony while many other manufacturers put their hopes on the VHS format. The difference is in recording techniques and the actual physical size of the video tape cartridge. VHS VCRs cannot play Beta-format video tapes and vice-versa.

At this time of writing, VHS has become the dominant recording technique used in the United States. There are still many Beta tapes around in the form of prerecorded movies but even Sony is now producing VHS equipment. In our discussion of video cassette recorders, therefore, we'll be talking primarily about VHS equipment (see Figure 6-1).

One of the best aspects of Beta recording was that the picture quality was somewhat better than the VHS format. However, VHS manufacturers now have a new type of VHS machine called "Super VHS" which provides a much higher screen resolution. The older VHS picture provided about 160 to 240 screen lines, but the Super VHS increases this resolution to 440 screen lines! For best results, however, special monitors with a C (color) and Y (luminance) input may be required.

At this time of writing, not many Super VHS machines are available. Also, there are few, if any, movies recorded in the new Super VHS format. If you're looking for best quality, therefore, be aware that you may be buying a quality picture, but without much to play on it, at least for a short time!

Figure 6-1.
Typical VHS VCR
Courtesy of Radio Shack
A Division of Tandy Corporation

Basic VCR Operation

One of the most important things to realize about video cassette recorders is that they are not only recorders, but miniature televisions as well. About the only thing missing from the television portion of the VCR is the picture tube! Some people simply will not believe that a VCR contains a complete television tuner, but it does. You can record program material on one channel while watching another channel on your normal television.

Your television antenna normally attaches to your VCR ANT IN jack (also called IN FROM ANT or VHF/UHF INPUT), with an output from the VCR's ANT OUT jack (also called OUT TO TV or VHF/UHF OUTPUT). When not playing back material from a prerecorded video cassette, the antenna is routed to both the VCR electronics and the television, as shown in Figure 6-2. Since the signals on the antenna include all channels, the VCR can record one station while passing through the antenna signals for viewing any other station on the television.

During playback mode, however, the video cassette video and audio signal is routed to your television, where it is received on channel 3 or 4 (or through the audio and video jacks of a monitor). During playback mode, of course, the VCR cannot record, as there is only one video cassette that can be in the machine at one time.

Like a television, all VCRs will tune to any VHF channels (2 through 13) or UHF channels (13 through 83). Some will also tune to cable tv channels (see below).

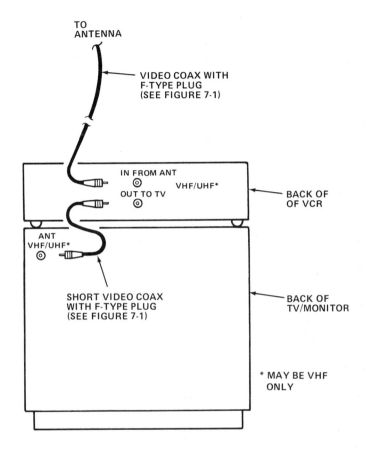

Figure 6-2.
VCR Antenna Connections

Almost all VHS recorders have three tape speeds - SP, LP, and EP (standard, long, and extended play) - allowing you to record two, four, or six hours of material on standard T-120 video cassette tapes. However, picture quality is degraded in LP and EP. In these modes, the picture resolution is not good, and images tend to look "blocky", without detail.

VCR Buying Guide

Here are the specifications and buzzwords to watch out for when buying a VCR.

Two Versus Four Heads

Most VCRs have two recording heads. However, some of the more expensive models have four heads. The primary reason for the four heads is to make the image less "glitchy" without horizontal "tear" lines and snow in the forward or reverse scan modes. (In forward and reverse scan, the picture is shown at 10 times or 15 times the normal rate so that you can search forward or backwards). In four head VCRs the quality of this picture is much better.

Analog Versus Digital VCRs

Older VCRs are *analog* sets. This means that the video is processed as a continuous series of images. *Digital* VCRs are a new type of set. They came on the scene about the same time as digital televisions. Digital VCRs process video as a series of digital pictures in much the same fashion as CD players break up an audio signal into a series of ones and zeroes.

There are several advantages to digital sets. One of these advantages is *picture-in-picture* (PIP). Picture-in-picture capability is also present in digital televisions. In this mode, the normal picture from VCR playback or from an off-the-air television program is displayed on the television connected to the VCR. Within this picture, say at the lower right-hand corner, is an image from another channel, displayed at 1/4 or 1/8 size. You can, therefore, be playing back the movie "Friday the 13th, Part 16" on your digital VCR, and at the same time watch a tiny image of the Sunday football game as well! (See Figure 5-3).

Digital sets in forward and reverse search modes display a glitch-free picture without additional heads. Digital sets also have the ability to *freeze* a frame from the VCR cassette material or an off-the air picture. This is helpful for stopping action during sports events or in freezing a telephone number momentarily displayed on the screen. *Slow motion* is also a feature on some digital sets.

Picture Quality

Chances are, if you go into a showroom and ask a clerk what the video frequency response of a VCR is, you'll get a blank look. However, this is an important specification. Generally, most VCRs have about a 2 to 3 megaHertz frequency response, corresponding to about 160 to 240 lines on the television picture (each one megaHertz corresponds to about 80 lines on the screen). The higher this frequency response is, the better. Super VHS and Beta will have better frequency response than standard VHS.

Newer VCRs (from about 1987 on) have an "HQ" (high quality) system for improved picture quality. This feature is a set of electronic circuit design options which truly does improve picture quality, reducing noise and "snow". The HQ system is usually not an option, but comes free with the newer sets, and is a desirable feature to have.

Comb filters, an electronic circuit also found in television sets, are found on better VCRs. Comb filters provide extended frequency response for video (higher resolution pictures).

Tuner

The tuner on a VCR serves the same purpose as a tuner on a television. It tunes to a VHF or UHF channel. Newer VCR tuners are "cable ready". They will not only tune to a VHF or UHF channel, but will tune to cable channels as well, without having to mechanically tune to the cable "subchannels". (Older televisions and VCRs cannot receive some of the cable television subchannels.) This means in some cases that you will not have to rent a decoder box from the cable television company serving your area. (Scrambled subscriber services, however, like HBO or The Movie Channel will require the decoder box to descramble the signal - the VCR tuner will not do any descrambling).

Newer tuners are digital tuners that do not have any mechanical switching to change channels. They also use a type of tuning called *frequency synthesis* (quartz tuning), which simply means that the tuning is done electronically.

Programming

All VCRs have a built-in clock that displays on a front panel display (usually liquid-crystal) and can be used to *program* recording. You could, for example, set the VCR to automatically turn on at 8:00 pm to record channel 2. In newer sets, you can do this 14 to 28 days in advance for four or eight separate programs. One set even allows you to program up to one year in advance!

This programming ability is a handy feature to have, as you can "time-shift" television programming by recording programs while you are away and play them back at your convenience.

One fallacy in recording multiple programs is that you'll probably be playing the tape in fast SP mode for best picture quality. A T-120 video cassette tape allows only two hours of recording in this mode, so you're probably limited to two one-hour programs, anyway.

Newer VCRs have an *on-screen programming* capability, as shown in Figure 6-3. The television screen displays a step-by-step guide to setting the program times via printed text. You can reply via the remote control keypad. One of the problems with VCRs is that they provide too many options, and programming seems unnecessarily complicated at times. The on-screen programming helps.

Tape Counter

All VCRs have a built-in *tape counter*. On earlier sets this was a mechanical device, but now it's invariably electronic. The tape counter counts forward or backwards to find the 0000 setting on a tape. Typically, you'll want to set the start of the tape to 0000, or reset the counter at a point on the tape to which you want to return.

```
PRG NO.2 WEEKLY
START DAY  03/28/88
                (MON)
START        STOP
09:30 PM   >10:00 PM
CHANNEL       5
      CANCEL?
      YES  >(ENTER)
      NO   >(MENU)
```

Figure 6-3.
On-Screen Programming

MTS Stereo and SAP

Some VCRs have a *stereo* capability. With this feature, you can record programs in stereo. You can then playback your own stereo tapes or commercial movies. Not all broadcast material is in stereo, but predictions are that most of it will be shortly.

The interesting thing about VCR stereo recording is that its quality is almost as good as CD (compact disk) sound recording when a special *high-fidelity* mode (available in newer VCRs) is used. Many people have used VCR tapes for audio alone, as tapes are very inexpensive and provide about six hours of listening. At this time of writing there is a great deal of controversy about DAT (Digital Audio Tape recording). DAT recorders provide extremely high-quality tape recording. Recording companies are afraid that DAT will be used to pirate copies of CD disks. However, VCR stereo in high-fidelity mode is an existing form of very high quality audio tape with flat response from about 20 to 20,000 Hertz.

Dolby stereo (see Chapter 2) is also offered on VCRs. Dolby stereo reduces tape "hiss" for quiet passages.

Some VCRs also offer the ability to record stereo from *simulcasts*. Simulcasts use one channel of audio from a television channel and the other channel from a simultaneous broadcast over an FM station. With widespread use of stereo telecasting, however, simulcasts may be a thing of the past.

Finally, some VCRs provide a decoding of *SAP* (second audio program) material. SAP is typically a second language broadcast superimposed over the normal sound and decoded by some televisions or VCRs.

Remote Devices

Many VCRs offer a *remote control* selection. Older VCRs used a wire leading to the VCR, but all newer sets have an infrared remote control device, which transmits tuning information on an invisible infrared light beam.

Better VCRs have a *direct* entry remote in which the numbers of the channel can be entered by keys on the remote. Older models used buttons to "count up" or "count down" channels until the proper one was reached. (This can be tedious when trying to enter channel 60 from channel number 2!)

Most remote control devices provide many of the same functions found on the front panel of the VCR. You can, for example, turn the VCR on and off, record, do a fast search forward, change the channel or programming, turn down the sound, and so forth. Some remote control devices will also work with the television as well, especially those from the same manufacturer. Remotes are advertised as "33-function" or "45-function" remotes. Generally, the more functions available, the better, even though you may never use some of them!

VCR Players

Some mention should be made of VCR *players*. These are VCRs that will only playback prerecorded cassettes, and will not record. The price difference between such players and a complete VCR recorder/playback unit is often not that great. If you watch only prerecorded material, the player is probably worthwhile, otherwise the total VCR is a better buy.

VCR Prices

Current prices on VCRs range from under $200 for a monaural (non-stereo), two-head VCR with remote, to over $600 for an MTS stereo, digital, model with on-screen programming. Often, the "guts" of the VCR are not that much different between the inexpensive model and ones that cost almost twice as much. As in other electronic equipment, often you are paying for the "intelligence" provided by digital signal processing and built-in computer circuitry. However, some of these features that you're paying for are very desirable - if you're a sports buff, for example, who can put a price tag on slow motion capability? The choice is yours!

Camcorders

A *camcorder* (Figure 6-4) is a combination video camera and video cassette recorder. It is a replacement for a home movie camera and offers

Figure 6-4.
Typical Camcorder
Courtesy of Radio Shack
A Division of Tandy Corporation

some distinct advantages. For one thing, there is no film to be developed. You can immediately play back the recorded video image and sound on a standard VCR (or on a television monitor). Another feature is that you have a built-in sound capability, which isn't always available on a movie camera. Finally, it's easier to *edit* the resulting video -there's no film to splice, as splicing can be done electronically.

All of this versatility doesn't come inexpensively, however. Current camcorders are priced from about $950 to $2000 or more.

There are three basic types of camcorders available today. The first type uses standard VHS tape. The second type uses a smaller, more compact VHS tape format called VHS-C, about the size of a pack of cigarettes. The third type, promoted by Sony, uses an 8 millimeter (8 mm) cassette tape format.

Taking pictures is easy with the camcorders. Load the tape cassette, frame the picture in a tiny black and white monitor that you see through an eyepiece, and start shooting. Depending upon the tape format used, you can record from one to two hours of material on a single cassette tape. During the course of the recording, you can rewind the tape, do a fast-search forward or reverse, freeze frame (some models), and do other things you would do with a standard VCR. When you are done shooting, you can rewind the tape, connect a cable from the camcorder to a monitor or television, and playback the action you've recorded.

Camcorder Buying Guide

Here are some of the specifications used to describe camcorders and some of the buzzwords to watch for.

Optics

Every camcorder has a *lens system*, which is similar to that of a film camera. Less expensive camcorders have a *fixed focus* lens that operates in much the same way as a film camera - point and shoot, but don't get too close to the subject. More elaborate systems have a *zoom* lens electrically powered by a single button. In some systems you can focus up to a few inches away from the camcorder, or immediately zoom out to a wider field of view. Many camcorders have *wide-angle* or *telephoto* lens capability. Typical lenses provide a 6 to 1 or 8 to 1 zoom capability. As in a film camera, the better and bigger the lens, the better will be the picture quality.

Many systems also have *infrared self-focusing*. In this type of system, the camcorder bounces infrared light rays off the subject and then uses the reflected echo to automatically adjust the focus. Also available on some systems is a *macro* capability for extreme closeups.

Some camcorders use an older style photo-sensitive picture element. Newer models may use a *CCD* (Charge Coupled Device) picture element chip. The CCD sensor may require more illumination (see below), but provides better resistance to "light streaking" and good color purity. Light streaking occurs when a camcorder is "panned" across a high-intensity light source - the light source leaves a visible trail on the picture.

Illumination

Like a film camera, one of the most important features is how little light is required to take a photograph. This capability is measured in *minimum illumination* expressed in *lux*. The lower the lux figure, the less light required to take a usable picture. Minimum illumination varies from about 8 to 22 lux for most camcorders. Some camcorders, however, permit recording down to 1 lux; these are very low light level cameras - one lux is the light of a single candle at 10 feet. Camcorders with CCD (Charge-Coupled Device) image sensors generally require more light for recording.

The *iris* (shutter) of the lenses is automatically adjustable on most cameras, to compensate for varying levels of illumination.

Another specification to look for on camcorders is the *white balance*, measured in IRE. The lower the IRE, the better the light balance (0 is perfect white balance). White balance refers to the amount of light appearing on neutral objects. Many camcorders allow you to set the white balance for indoor or outdoor shots, since the color makeup of each is different.

Resolution

Like television sets and VCRs, the resolution of a camcorder is a measure of how many lines can be displayed on a television screen. Consumer camcorders are not broadcast quality, and the horizontal resolution varies from 200 to 300 lines. The higher the number of lines, the better resolution that will be available. The resolution of the picture is really dependent upon the quality of the recorder portion of the camcorder, which is just a miniaturized VCR.

Super VHS camcorders are also becoming available. These camcorders record at the same resolution as Super VHS VCRs - about 480 lines - and offer excellent picture quality at a higher price. Like larger VCRs, HQ electronic circuitry is offered in some models, which improves picture quality.

Audio

All camcorders have a built-in *microphone* attached to the front of the unit. Most units provide monaural (single channel) recording, but more expensive units may offer stereo as well. Usually some provision is made to bypass the built-in microphone with another microphone input. This is handy for better quality recording where the microphone can be fixed in one place.

Recording Time, Battery Use, and Weight

Recording time on VHS format camcorders is two hours on SP and six hours on EP. On the smaller VHS-C format models, however, recording time is only one third that of VHS format - 20 minutes on SP.

All camcorders come with a built-in *battery pack* that typically allows use of the camcorder for about an hour before the batteries have to be recharged. Optional battery packs for longer periods of time are usually available. Recharging time is typically an hour. One thing to consider when using a camcorder on vacations: will the camcorder work on foreign voltages and frequencies? Common overseas voltages are 220 volts ac and 50 cycles per second rather than the United States 110 volts ac at 60 cycles per second. Look for a *multi-voltage AC adapter pack*. Also available for most camcorders is an a/c adapter to enable the unit to be plugged into a wall outlet.

As you might suspect, the 8mm and VHS-C camcorders come in smaller, more convenient packages and lighter weights, typically two to five pounds. The VHS versions are physically larger and may weigh seven- ten pounds. Even though the weight of the VHS model may not sound like much, it is while clambering over Mayan ruins in the Yucatan!

Chapter 7
Video Connectors,
Cables, and Options

There are only a few types of video connectors and cables but many different combinations, designed to fit different equipment. We'll describe the connectors, cables, and combinations in this chapter.

Video Connectors and Cable

Video signals are all very low-voltage signals which are safe to handle. There's also not much chance of harming equipment by plugging in the wrong cable.

Coaxial Cable

Most video cables are of the *coaxial* type. A coaxial cable is pictured in Figure 7-1. It has a center conductor, an insulating material around the conductor, and an outside *shield* made of woven metal fibers. Coaxial cables used for video gear are called *75-ohm coax cable for video use.* There are various types of jacks (connecting plugs) on the equipment. We'll talk about the most common types below. The important thing to remember is that almost any jack can be matched to any plug by an *adapter*, a small plug that converts one type of fitting to another (see Figure 7-2).

Coaxial cables can be run over dozens of feet and are available in 6 foot, 10 foot, and other lengths. You can even make your own cables by buying bulk coaxial cable, the proper plugs, and a *crimping tool* to attach the plug. For some connectors, no crimping tool is required - the plug just screws on without extensive preparation.

Twin-Lead

Another type of cable used for antenna systems is *300-ohm twin-lead*, shown in Figure 7-3. Twin lead is a flat, two-conductor cable, usually equipped with *spade lugs* on one end. It often comes from a television antenna. Twin-lead can be converted to a coaxial cable fitting and vice versa.

Plugs, Jacks, Males, and Females

It helps to use the proper term when you're buying a connector. A *plug* has a protruding center pin (or several pins in the case of multi-wire connectors). Plugs often are on the ends of cables. Plugs are also called *male* connectors.

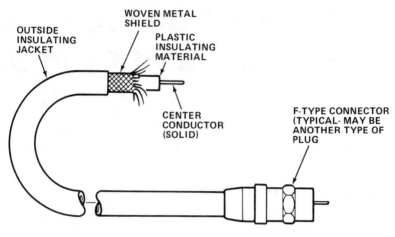

Figure 7-1.
Video Coaxial Cable

A *jack* is often mounted on the chassis of a piece of video gear. It has a receptacle that the plug fits into. Jacks are also called *female* connectors.

Often connectors are called by their male/female configuration followed by the type of connector. You'll see listings in catalogs for a "BNC male" or a "female F" connector.

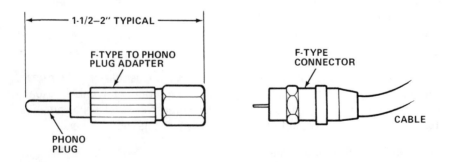

Figure 7-2.
Cable Adapter

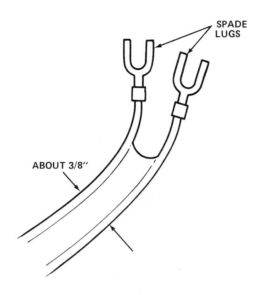

Figure 7-3.
Twin-Lead

RCA Phono Plugs and Jacks

One of the most common types of connectors is an *RCA* or *phono plug*. The phono plug is shown in Figure 7-4. It's used for both audio and video applications. When used for video applications, though, it is used on video coaxial cable, and not audio cable. Phono jacks and plugs (female and male, respectively) are used to connect to the video inputs and outputs of monitors, VCRs, camcorders, and other equipment. Phono plugs and jacks simply push on, without screwing.

F-Type Connectors

Another common type of connector is called an *F-Type* connector, shown in Figure 7-5. This is a threaded connector that screws on. F-Type connectors are used on cable television converter boxes, antennas, switch boxes, and generally equipment that connects antenna signals. A *push-on* connector replaces an F-type connector, and can simply be pushed on or pulled off.

BNC-Type Connectors

BNC connectors are not used as often as phono plugs and jacks or F-type connectors, but you may occasionally find them on professional quality equipment. They are shown in Figure 7-6. They fit in what is called a *bayonet* action - the connector is inserted over a guiding pin and then rotated to make the connection.

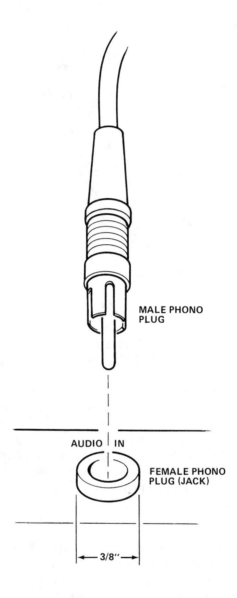

Figure 7-4.
Phono Plug and Jack

Adapters for Video Connectors

The three common types of video connectors - phono, F-type, and BNC-type - can be converted to one of the other types by inexpensive adapters available at Radio Shack stores. A female can be converted to a male and vice versa.

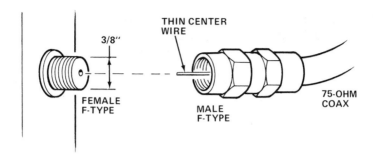

Figure 7-5.
F-Type Connectors

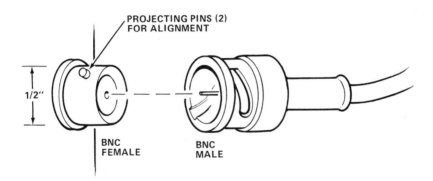

Figure 7-6.
BNC Connector

A 300-ohm coaxial cable can be converted to a 75-ohm coaxial cable fitting by a device called a *matching transformer*, commonly available, as shown in Figure 4-7. This converts an F-type fitting to spade lugs, or vice versa.

Audio Phono Jacks

The audio output from VCRs, camcorders, televisions, and other equipment is connected with the same type of RCA phono plug used for video. Audio cable instead of 75-ohm video cable can be used for audio connections. Cables in pairs are available, for the two channels of audio required in stereo equipment.

Phone Plugs

Some types of audio equipment use *phone plugs* (phone and not phono) and *jacks*, shown in Figure 7-7. There are two basic types of stereo phone plugs, the quarter-inch type and the *mini* plug. Both male and female versions are provided, although often the female type is on the equipment, and you'll need a cable with a male (plug) to connect the audio signal. Note that these connectors are *stereo connectors* (two insulating bands) and not *monaural connectors*. However, if you have monaural equipment to be connected, you can obtain monaural to stereo (or vice versa) adapters from any Radio Shack store.

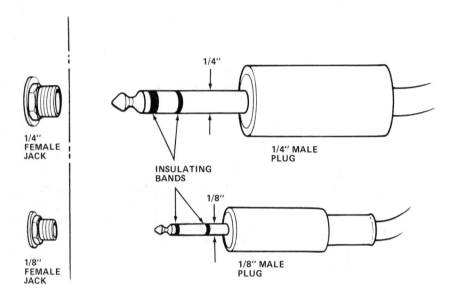

Figure 7-7.
Phone Plugs and Jacks

Audio/Video Cable

Cables labelled audio/video coaxial cable are good for either audio or video use. Usually this type of cable comes with phono jacks on either end. The phono jacks can be converted to other types of connectors with adapters. However, this cable should not be used in the antenna portion of a system because it will cause the signal to deteriorate (use 75-ohm coaxial cable or 300-ohm twin-lead instead).

Television Antenna Connections

Television Alone

Simple television antenna connections are shown in Figure 7-8. In fringe areas, you may require two antennas - one for VHF and one for UHF television

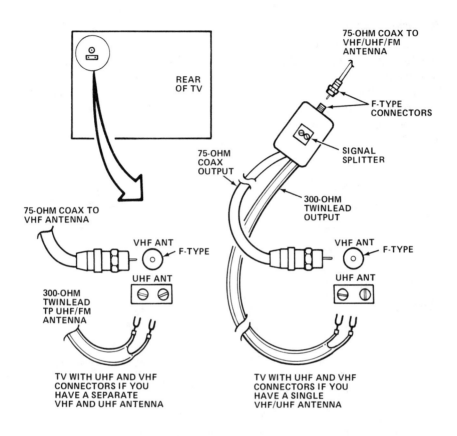

75-OHM COAX TO
VHF/UHF/FM
ANTENNA

REAR
OF TV

F-TYPE
CONNECTORS

75-OHM
COAX
OUTPUT

SIGNAL
SPLITTER

300-OHM
TWINLEAD
OUTPUT

75-OHM COAX TO
VHF ANTENNA

VHF ANT
F-TYPE

VHF ANT
F-TYPE

UHF ANT

UHF ANT

300-OHM
TWINLEAD
TP UHF/FM
ANTENNA

TV WITH UHF AND VHF
CONNECTORS IF YOU
HAVE A SEPARATE
VHF AND UHF ANTENNA

TV WITH UHF AND VHF
CONNECTORS IF YOU
HAVE A SINGLE
VHF/UHF ANTENNA

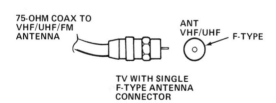

75-OHM COAX TO
VHF/UHF/FM
ANTENNA

ANT
VHF/UHF   F-TYPE

TV WITH SINGLE
F-TYPE ANTENNA
CONNECTOR

Figure 7-8.
Simple Tv Antenna Connections

channels. However, another more common option is a combination VHF/ UHF/FM antenna. In this antenna, a single download comes from the antenna. The signal is then split electrically into three outputs - one for VHF, one for UHF, and one for FM. You may need to add a matching transformer to convert the spade lug connectors to F-type connectors that fit your television.

Television with VCR

Antenna connections with a VCR are shown in Figure 7-9. In this case the antenna connections go to the VCR VHF and UHF terminals. You'll need to add additional 75-ohm coax or 300-ohm twin-lead to then go from the VCR terminals to the television terminals. You might also need one or two matching transformers. These connections enable the VCR and television to receive the same off-the-air signal when the VCR is not playing back, and yet route the VCR to the television antenna inputs for playback.

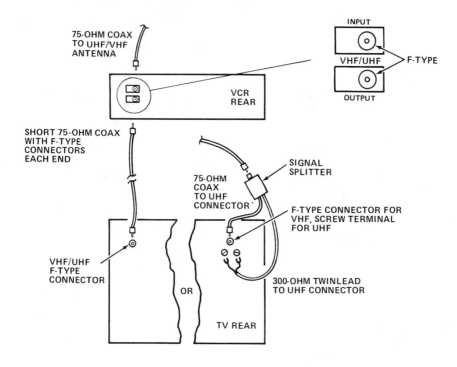

Figure 7-9.
Antenna with VCR and Television

Television and Cable Television Service

If your television is cable ready, it may not require a cable converter box. In this case, the F-type connector found in most cable systems can be connected directly to the VHF antenna connector on your television, as shown in Figure 7-10. Extend the cable with a length of 75-ohm coax with F-type connectors and an F-type male-to-male adapter.

If your television is not cable-ready, or if you are subscribing to movie services such as HBO, the cable from the cable system has to be routed to the

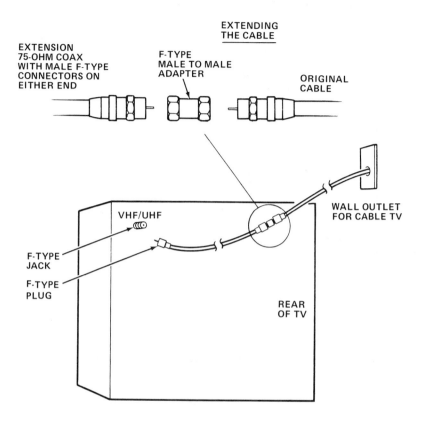

Figure 7-10.
Cable Television Connections to
Cable-Ready Set

cable decoder box. Usually, this has an INPUT F-type connector, and an OUTPUT F-type connector. A short cable with F-type connectors on both ends should be routed from the converter box to the VHF antenna connector on your television, as shown in Figure 7-11.

If your television *is* cable ready, switch to the CATV mode.

Television, VCR, and Cable Service

The connections here are similar to the television/cable converter box connections in Figure 7-11, except that the output of the converter box (or the cable from the wall) is connected first to the ANT IN connector on the VCR, and a short cable is run from the VCR ANT OUT connector to the television (see Figure 7-12). Both the television and VCR should be switched to the CATV mode.

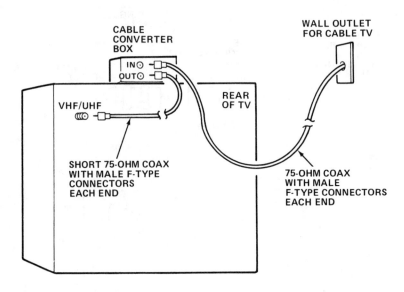

Figure 7-11.
Cable Decoder Box to Tv Connections

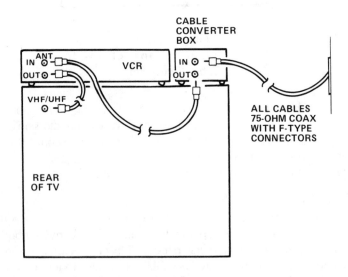

Figure 7-12.
Cable Decoder, VCR, and TV
Connections

Monitor Connections

If your television is a *monitor* as well as a television receiver, you can get a better quality picture by connecting the VCR VIDEO OUT, usually a phono jack, to the VIDEO IN jack on the television. This input is then selected by a VIDEO SOURCE switch. Also connect the VCR AUDIO OUT (one or two jacks) to the AUDIO IN jacks on the monitor. The connection is shown in Figure 7-13. Change the tv/monitor input to AUX or EXTERNAL for viewing.

Audio Connections

The audio output from the VCR or tv can be connected to a sound system. If your VCR is a stereo model, you can get complete stereo sound, both on prerecorded stereo video cassettes and on off-the-air programming in stereo. The basic connection is from the VCR AUDIO OUT jacks (one for a monaural VCR, two for a stereo VCR) to the AUX or TV IN jacks of your amplifier or receiver/amplifier in your audio system. This is the only link that you will need for audio - it's pictured in Figure 7-14. Turn off your television sound and control the sound output via your stereo system for this setup. The channel sound selected will be the one displayed on your VCR.

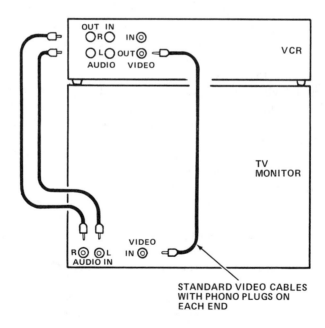

Figure 7-13.
Monitor Connections

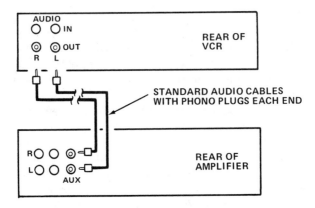

Figure 7-14.
Tv and Stereo System Connections

Splitter Configurations

Some cable converter boxes degrade the audio or video signal. Sound especially sometimes suffers in the decoding process. Also, the output of the cable converter box is often on channel 2 or 3, meaning that the television or VCR must be set to that channel, with switching done via the cable converter box. To switch from HBO to channel 2 requires switching the converter box tuner setting from 23 to 2, for example.

You may want to *bypass* the converter box to get better quality video and audio if your television and/or VCR is cable ready. This can be done with a *signal splitter*. A signal splitter a small box that splits the antenna signal into two or more signals, as shown in Figure 7-15. Typically there is one F-type input and two F-type outputs.

With the layout shown in Figure 7-16, you can split the signal from the cable before the cable converter box and run it to both the cable converter box and the VCR. The VCR can then record on any channel (by setting the channel on the VCR) while you're watching another channel on your television (on channel 2 or 3 from the converter box). The input to the tv is switched by an *A/B switch* to either VCR output or cable converter box output. Add a second A/B switch, and you can switch the input to the VCR from either the cable input or the output of the converter box. The A/B switch is a tv switch that simply selects one of two inputs, as shown in Figure 7-17.

One thing to be concerned about: The signal splitter does cause some loss of signal when inserted in an antenna connection. One correction for this is to use a *signal amplifier*, as shown in Figure 7-18. The signal amplifier should be a type either for an antenna or for a cable system. It boosts the entire range of signals present on the antenna or cable, and compensates for any loss of signal from cable splitters.

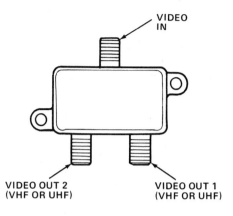

Figure 7-15.
Signal Splitter

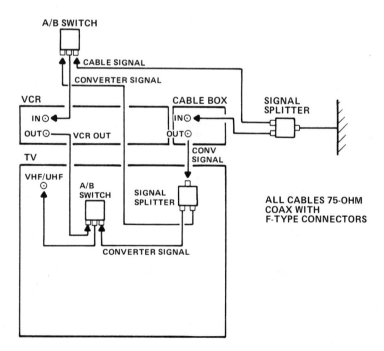

Figure 7-16.
Simultaneous Cable and Converter Box
Input

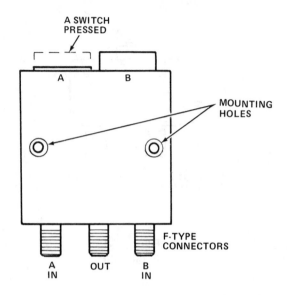

Figure 7-17.
A/B Switch

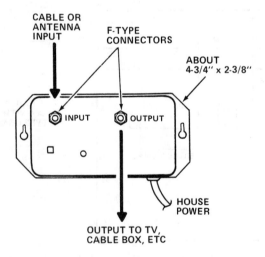

Figure 7-18.
Signal Amplifier

More Complicated Video Connections

As you can see from the above connection diagrams, tv connections can be complicated and require some additional equipment. One solution is to add a *video selector*, as shown in Figure 7-19. The video selector is basically a video switch box, which allows you to switch between a tv antenna, cable input, VCR input, and one or two auxiliary inputs. You might have a video disk player, for example, as one of the auxiliary inputs. On the output side, outputs go to a primary tv, second tv, or VCR input. Output and input connections are usually F-type connectors.

When making these connections, it helps to have a number of cables of varying length with male F-type connectors on each end, or better yet, *quick connect* connectors. Work a step at a time, connecting first the antenna and cable inputs and then testing to see whether the signal routing is correct.

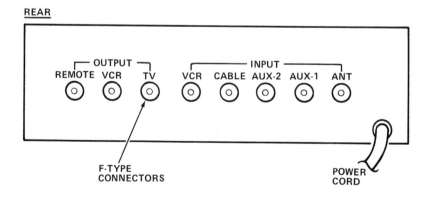

Figure 7-19.
Video Selector

In more elaborate connections, it also probably pays to invest in a *crimping* tool and individual connectors so that you can custom tailor the cables to convenient lengths. This avoids many coiled cables that can't be stored easily behind equipment.

Remember that with these connections you'll be working with only the antenna inputs of the television and VCR, and not with the VIDEO OUT or AUDIO connections.

Copying From One VCR to Another

It is possible to copy video tapes from one VCR to another. The connections are shown in Figure 7-20. Just run a phono jack cable from the VIDEO OUT connector on the "source" VCR to the VIDEO IN connector on the "destination" VCR and another one or two phono jack cables from the AUDIO OUT (source) to AUDIO IN (destination). The source VCR may also be a laser disk player or other video source, such as the early CED video disk players. The source VCR should be the higher quality VCR of the two - recording from a Beta VCR works better than from a VHS VCR, for example, because of the better picture quality of many Beta VCRs. You will have to select EXTERNAL or AUX on the VCR which is doing the recording to choose the VIDEO and AUDIO inputs.

Some prerecorded tapes have a *Copyguard* protection system to defeat rerecording of copyrighted material. The result will not be acceptable when copied to another video cassette. (See the *video processor* below.)

Other Video Equipment

In addition to the devices for splitting and amplifying antenna signals above, there's a whole host of other video equipment to switch, enhance

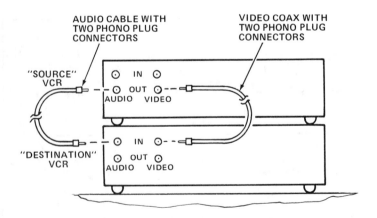

Figure 7-20.
Copying from One VCR to Another

picture quality, and provide MTS sound. We'll mention some of them here.

MTS Stereo Converters

You can add MTS stereo sound to your present tv system by adding an MTS stereo receiver, shown in Figure 7-21. This is like an FM receiver, but receives the sound portion of VHF and UHF tv channels, decoding the sound into two channels. The two channels of sound can then be routed to a stereo VCR. The receiver comes with a built-in stereo amplifier and speakers, or you can route the audio output to your sound system.

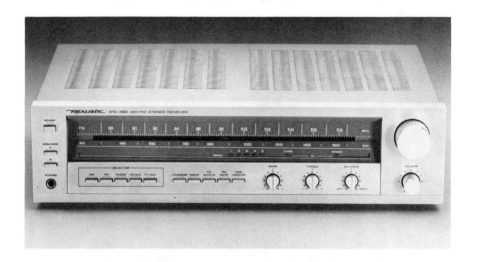

Figure 7-21.
MTS Stereo Receiver
Courtesy of Radio Shack
A Division of Tandy Corporation

Video Processor

This piece of video equipment (Figure 7-22) is a picture enhancer, reducing roll and jitter, eliminating some "snow" and in general reducing noise in VCR taping. Two video sources can be selected as inputs. This model also has an RF output so that a normal tv (not a monitor) can be used with a VCR.

Video Distribution Systems

Video distribution systems (Figure 7-23) allow you to send a VCR, antenna signal, or cable signal to multiple televisions located in different parts of your house. The connection to each tv is through a single video cable, eliminating any complicated switching. This means that you can still use your video system and all of its equipment at one location and watch different channels in three other locations.

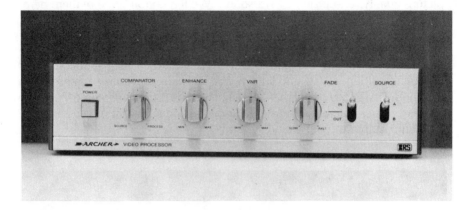

Figure 7-22.
Video Processor
Courtesy of Radio Shack
A Division of Tandy Corporation

Figure 7-23.
Video Distribution System
Courtesy of Radio Shack
A Division of Tandy Corporation

Chapter 8
Security Systems and Home Controls

Crime statistics are up! There are more robberies and thefts than ever before. Using security systems, however, you can avoid theft problems before they occur. You can also protect your home and office from fire. In this chapter we'll take a look at devices that protect you, both from intruders and fire and smoke. We'll give you buying tips on these devices and also provide information on how to hook them up. We'll then look at appliance and light controllers, devices that enable you to control your lights and appliances automatically.

Alarm Center with Wiring

General

A diagram of a full-blown burglar and fire alarm is shown in Figure 8-1. It is a complete system with just about every type of protection available. Your own system can be as elaborate as this, or may be scaled down for your own requirements and budget. We'll describe the system in general terms and then talk about each individual component of the system in detail.

The system uses a central burglar and fire alarm center as an "alarm center". This is a medium-sized box, similar to a circuit breaker box, into which all wires of the system are connected.

Coming into the alarm center are wires from individual sensors located in all the rooms of the house. These wires are low-voltage wires that are not dangerous to handle. The wires are strung through the attic, under the foundation of the house, through walls or ducting, tacked to baseboards or under carpeting, or wherever convenient routes to the rooms of the house can be found.

Depending upon the house and requirements, wires may connect to these devices:

- A magnetic switch mounted in a door or window frame. If the door or window is opened while the system is "armed", it will signal the alarm center of an intrusion.

- Foil tape laid across window glass. If the window is broken, the alarm center will be signaled.

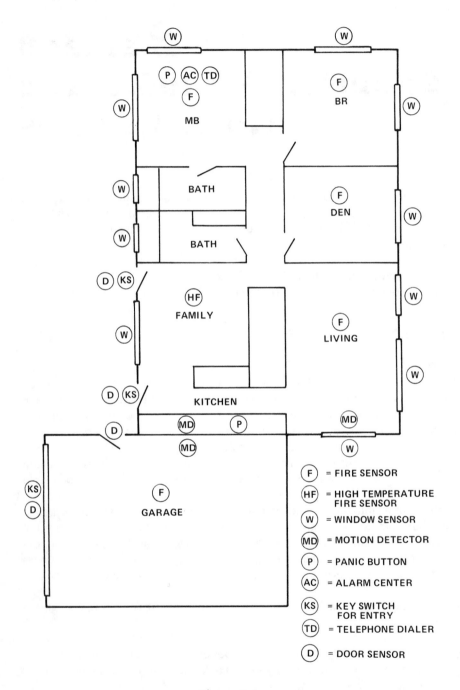

Figure 8-1.
Complete Security System

- Vibration detectors on windows. If the window is broken or opened, the alarm center will be signaled.

- Vibration detectors put on certain expensive objects, for example, automobiles or works of art.

- An infrared sensor that detects an invisible infrared light beam across a room or doorway. If the beam is broken, the alarm center is signaled.

- An infrared or ultrasonic motion detector. Operating by invisible infrared light or sound waves beyond human hearing, these devices detect any movement within a room and signal the alarm center.

- Smoke and heat sensors. These devices are similar to individual smoke alarms, but in addition to sounding a loud alarm, signal the alarm center.

- Special high-temperature fire alarms. These alarms are located in areas where normal activity would trigger standard smoke alarms, such as kitchens. Again, they signal the alarm center.

- Panic alarms. Located throughout the house, these buttons can be pushed to signal the alarm center. A special small panic alarm can be worn around the neck and activated by pushing a button in an emergency.

- Pressure sensitive pads. These pads are laid upon the floor at entrances and are activated by someone walking over them. The alarm center is signalled by the pressure switch contained in the pads.

The alarm center continuously monitors these sensors through their wired connections. If any of the sensors are triggered, either by intrusion, vibration, motion, or fire or smoke, the alarm center instantaneously receives a signal and then sounds an indoor and outdoor siren. In addition, the system automatically dials three preset telephone numbers. One of these numbers is to a nearby neighbor. One of the numbers is to your office. The last number is to a relative. For each number dialed, a prerecorded message is played.

The entire system has battery backup, so that it is fully functional during ac power failures.

The system is turned on and off by a security switch. For example, before retiring for the evening, you enter a four-digit secret code into a digital key switch, after checking all windows, doors, and sensors. The system delays a minute or so to give you time to exit a room in which there is a motion detector, and then becomes armed. On leaving the house, the system is armed by a second digital key switch.

An Alternative Approach

The system described above is an inexpensive system relying upon low voltage wiring strung throughout the house. Another type of system uses the same type of sensors - magnetic switches and the like - but uses several transmitters. The transmitters are actually tiny radio transmitters that signal a alarm center. More than one sensor can be connected to one transmitter. For example, you could have a transmitter in one room with five sensors connected to it.

This approach works well, but tends to be more expensive than the system described above. Since the same type of sensors are used with this system, the descriptions below will apply to both types of systems, with the exception of the description of the alarm center.

The Alarm Center

The alarm center contains a control panel similar to the one shown in Figure 8-2. Indicator lights indicate whether circuits are activated or not activated. Switches enable or disable individual circuits. Other buttons test the circuits.

The heart of the alarm center is a series of screw terminals that are used to connect individual wires leading to sensors, to connect alarm bells, and to connect locks and key switches.

There are two basic types of circuits that can be connected, *normally open* (NO) and *normally closed* (NC) circuits. Figure 8-3 shows the difference.

Normally open circuits connect to sensors that are open circuits. They are closed by intrusion or fire. Most fire sensors, for example, are NO sensors. NO sensors are connected *in parallel* with each sensor being

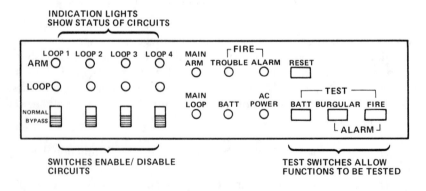

Figure 8-2.
Alarm Center Panel

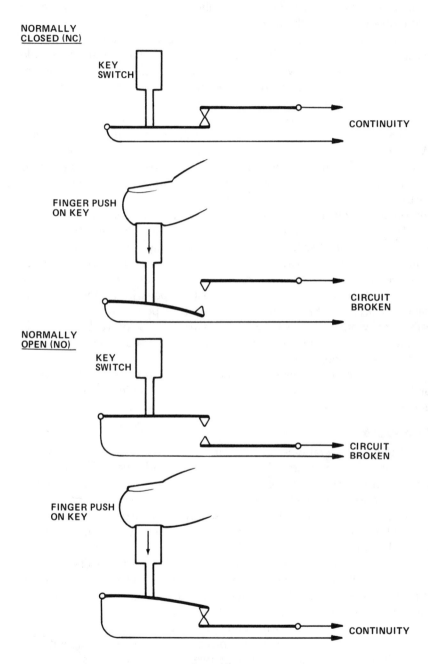

Figure 8-3.
Normally Open and Normally Closed
Switches

connected to the pair of wires leading to the alarm center circuit. Any sensor that closes will trigger the alarm center.

Normally closed circuits connect to sensors that are closed circuits. Most intrusion sensors are of this type because if an intruder tries to disable the sensor by cutting a wire, he will open the circuit, causing an alarm. Normally closed sensors are connected in series, so that activation of any one sensor will trigger the alarm center.

The alarm center has a provision for momentary contacts (normally open or closed) as well. For example, the panic alarm is a momentary alarm. When triggered by a neck panic button push, it takes only a momentary push to activate the alarm - the button does not have to be held down continually.

The alarm center also has a provision for "delayed" sensing. In these cases, there is a slight delay while the circuit is set or reset. This allows the sensors to be activated after people have left a room, for example.

Special connections connect to key switches which enable the system, to alarm sirens, and to a telephone dialer. There may also be a connection to a unit that connects to an ac plug so that any electrical appliance or light can be turned on automatically, for example, an outside light.

The alarm center has room for a battery pack inside the case. The battery pack powers the alarm center in the case of power failure. The system is therefore workable even if someone has turned off ac power by an outside circuit breaker. Batteries last from 20 minutes to four hours, depending upon the type of batteries used.

This type of alarm center is on the order of $100 to $150 for the basic unit, as opposed to about $300 for the radio transmitter type described above.

Fire Alarms

There are several types of fire alarms that may be used in a home. A *stand-alone fire alarm* detects smoke or fire by ionization of the air or by optical sensing. The ionization type contains a minute amount of radioactive material (posing no health risks). It detects not only smoke, but invisible products of combustion. A loud alarm sounds if smoke or fire is detected. A test button is used to test the alarm manually. This type of alarm cannot normally be connected to a security system.

An *alarm center fire alarm* (Figure 8-4) is designed to operate in conjunction with the alarm center. It not only sounds a loud alarm from the unit itself but closes a circuit to trigger the alarm center through wire connections. This type of fire alarm is not meant to be used stand-alone, as it requires power from the alarm center. These units cost about $35.

Special heat sensors are used in places where the room temperature is normally high, as in attics, or where there may be quantities of smoke, as in kitchens. Heat sensors are available for 135 and 190 degree temperatures. These sensors also close a circuit and can be used with the alarm center. These units cost about $7 each.

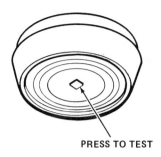

PRESS TO TEST

Figure 8-4.
Fire Alarm for Alarm Center

Many individual fire or heat sensors can be connected in a single circuit. Any sensor that is triggered will immediately cause the alarm center to activate an alarm.

Window and Door Sensors

There are many different types of window and door sensors, but they fall into the two general categories described above, normally open or normally closed. The NC is probably best as it is activated when cut.

Foil tape is used across windows, especially large windows. Breaking the window cuts the foil as well, opening the circuit. Another type of window sensor is the glass breakage detector. This sensor contains a small switch that is activated by vibration. When the window is broken or even opened, the sensor will be activated. The amount of vibration necessary to activate the switch can be adjusted, to prevent aircraft noise or other noise from triggering the circuit incorrectly. The foil tape and vibration sensors are shown in Figure 8-5.

Magnetic switches are in two sections. Each section contains magnets. When the magnets are close together the switch is closed or open. Moving a window sash or a door separates the two magnets and allows the switch to close. See Figure 8-6.

A *plunger* type of switch has a small plunger that keeps the switch closed or open. When a window or door edge is pulled away from the the corresponding edge, the spring-loaded plunger is released and activates the switch. See Figure 8-7.

Key Switches

Key switches come in two types. There is a switch that is set on or off by inserting and turning a small key. A second type of key switch is actually a *key pad*. The key pad has a push button telephone type of pad with 12 keys. The key switch can be set to any four digit code, allowing about 17 million

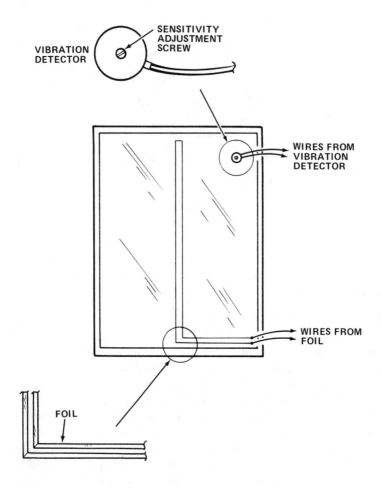

Figure 8-5.
Foil and Vibration Window Sensors

separate codes. The switch is activated by pressing the keys in sequence. LED indicator lights show when the circuit is armed or disarmed. Both types of switches are shown in Figure 8-8.

Motion Detectors

There are three types of motion detectors - *infrared, ultrasonic,* and *microwave.* Infrared motion detectors detect intruders by detecting body heat and motion. Ultrasonic motion detectors detect intruders by sending out high-frequency sound waves beyond the range of human hearing and detecting a change in their reflection. Microwave motion detectors use the same principal, but send out very high frequency radio waves.

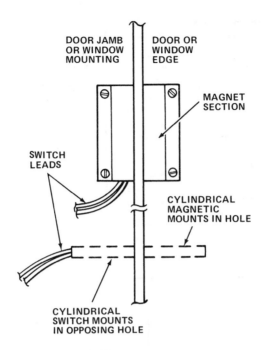

DOOR JAMB
OR WINDOW
MOUNTING

DOOR OR
WINDOW
EDGE

MAGNET
SECTION

SWITCH
LEADS

CYLINDRICAL
MAGNETIC
MOUNTS IN HOLE

CYLINDRICAL
SWITCH MOUNTS
IN OPPOSING HOLE

Figure 8-6.
Magnetic Switches

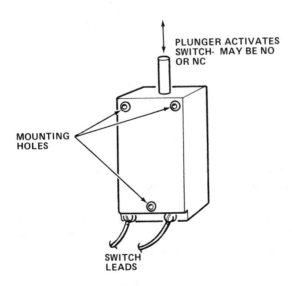

PLUNGER ACTIVATES
SWITCH- MAY BE NO
OR NC

MOUNTING
HOLES

SWITCH
LEADS

Figure 8-7.
Plunger Switch

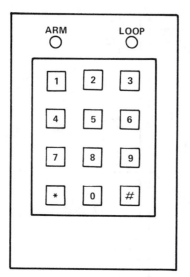

DIGITAL KEY SWITCH-
CODE ENTERED VIA
TELEPHONE-LIKE KEYPAD
TO ACTIVATE/ DEACTIVATE
CIRCUIT

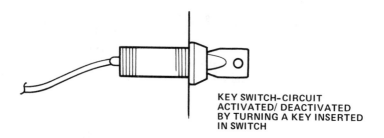

KEY SWITCH–CIRCUIT
ACTIVATED/ DEACTIVATED
BY TURNING A KEY INSERTED
IN SWITCH

Figure 8-8.
Key Switches

Typical coverage for motion detectors is an area 20 by 30 feet. This coverage is affected by the room shape and objects in the room.

Motion detectors have an *entry* and *exit* delay - the alarm does not sound for 15 seconds or so to provide time enough for the owner to walk in and turn off the alarm or to leave the room after the alarm has been set.

Motion detectors range from about $40 to $100. Most have NO or NC connections that enable them to be hooked up to an alarm center such as the one described above. A typical motion detector is about the size of a large book turned on its side; some have been disguised as books.

One popular type of motion detector is not designed to be connected to an alarm center. Rather, it has a built-in flood light or lights which are turned on for a variable period of time (10 seconds to 15 minutes) after motion is detected. This is an excellent device to use outside a house to replace an outside front or back porch light.

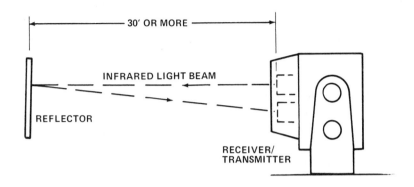

Figure 8-9.
Light Beam Device

Infrared Light Beam Devices

An infrared light beam device is an update to the old type of photodetector, in which a light beam shines across a doorway and is detected by a photocell. In this case, though, the light beam is of infrared light and is invisible. The receiver/transmitter unit is about the size of a small camera. It sends out an invisible infrared light beam which is reflected back by a *beam reflector* (see Figure 8-9). The distance covered can be as great as 30 feet or so, so the device can protect quite a wide area.

Most of these units have NO or NC connections which can connect to an alarm system. The cost is typically $35 to $70.

Emergency Phone Dialers

An *emergency phone dialer* is essentially a tape recorder that attaches to a phone line and then automatically dials one or more numbers. When the number is reached (the device can tell if there's a busy signal) a prerecorded message is sent. The message can be any message you care to record, such as "HELP! BREAK-IN AT 250 OAK LANE, TECHNOLOGYVILLE! PLEASE SEND HELP" or "FIRE AT 250 OAK LANE, TECHNOLOGYVILLE! PLEASE INFORM FIRE DEPARTMENT".

The phone dialer is activated by a switch closure from an alarm center or by simply closing or opening a detector switch. The device can therefore be used either in an alarm center setup or can be triggered by a motion detector or light beam interruption.

Typically, up to 20 phone digits can be programmed into the units, and they will work on pulse or tone telephone lines.

Many police and fire departments will not accept such calls, so it may be necessary to program the unit to call friends, relatives, or security services instead. Of course, the unit can also call you at a daytime or nighttime number, and when coupled with call forwarding offered by telephone companies can reach any number where you might be during any time of the day or night.

Telephone dialers range in price from about $100 to $200.

Connecting Alarm Systems

The basic alarm system concept is fairly simple. Since there are no high voltages present on wiring to sensors, ordinary *hookup wire* can be used with no fire or shock hazard. Just remember to stay away from house wiring when stringing the alarm system wire. Start with the basic alarm system connection to the ac circuit and a battery backup. Then add circuits one by one, testing each circuit as you add it. For example, if you're adding a fire alarm, actually blow some smoke towards the alarm to see if it activates properly.

Any number of NC sensors can be added in the same circuit or any number of NO sensors can be added in the same circuit. Logically group the circuits by the function (fire, intrusion sensors, or key switches). The hardest part of this is to string the wire throughout the house, one reason security companies charge so much for the initial installation ($1500 to $2000). However, with patience and some planning, you can install a better system at far less cost than a security system can provide.

AC Remote Control Devices

Another type of system we'd like to mention here is not necessarily a security system. Radio Shack calls the system *Plug 'N Power*. It consists of a central control unit with a built in clock that plugs into a wall socket. However, not only is the central control unit powered from a wall socket, but it communicates with other modules via *house wiring*. Here are some of the things the system will do:

- Turn a light off or on at any prearranged time, several times during the day.

- Turn an appliance on or off at any prearranged time, several times during the day.

- Dim certain lights at prearranged times.

- Allow you to control lights or appliances from your car or outside your home.

• Instantly turn on all house lights with a single switch from your bedside.

Using Plug 'N Power, you can program the lights in your home to turn on and off during vacations or during late evening hours when you're asleep. You can also start the coffee pot while you're still sleeping. After driving up in the driveway, you can turn on all the house lights. If you hear a noise late at night, you can immediately turn on all house lights.

The system works with Plug 'N Power modules, shown in Figure 8-10. One type of module plugs into a wall plug and then a lamp or appliance plugs into a plug on the module. Another type of module replaces a wall

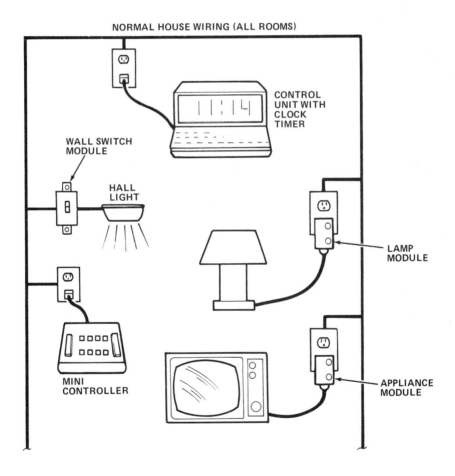

Figure 8-10.
Plug 'N Power System

switch with a wall switch module - the wall switch works normally, but is also controlled by the central control unit.

More than one control unit can also be used. The master central control unit can be supplemented by a second or third control unit in other rooms.

The big advantage of such a system is that it uses existing ac wiring. No wires need to be strung around the house. The central control units cost from about $13 to $40 and individual modules, each of which controls one lamp, appliance, or wall switch, cost about $13. A basic system controlling a front outside light, a rear outside light, a side outside light, lights in four rooms, and an appliance such as a coffee pot, will cost approximately $150.

Chapter 9
Telephone Equipment

After the famous Carterphone court decision in 1968, manufacturers other than telephone companies could connect equipment to telephone lines. As a result of this decision and inexpensive electronic components, there is now an abundance of telephones that can store and dial numbers, redial automatically, forward calls to other numbers, act as a speakerphone, answer the phone and record messages, and perform other magic tricks unthinkable 30 years ago.

In this chapter we'll survey existing telephone equipment and show you how to connect telephones to existing lines. It's easy!

Telephone Basics

Telephones operate with two wires, as shown in Figure 9-1. These wires carry voice in both directions and also carry a ringing pulse that rings the phone. This is probably a good time to mention that the ringing pulse is about 90 volts, enough to give you a mild shock if you happen to be touching both sides of the line while the ringing pulse occurs. If you have heart or other health problems, it's probably best to avoid working on active lines.

The sound carried by telephone lines is not a very good quality - the high and low frequencies are cut off so that speech sounds muffled. This is the nature of the phone system - it's not designed to be a high-fidelity network, as anyone who ever listened to "The Yellow Rose of Texas" while waiting for a sales clerk will attest!

Telephones dial numbers by two methods. The oldest of these is known as *rotary* or *pulsed* dialing. To dial the digit eight, for example, eight pulses are sent out over the lines, to dial four, four pulses are sent, and so forth. Each pulse corresponds to a "break" in the circuit. It's still possible to get the "0" operator by rapidly pressing the cradle button ten times in succession (breaking the circuit ten times), if you're coordinated enough.

The newer dialing method is *Touch-Tone dialing*, also known as *tone* dialing. In this method of dialing pressing a button on the phone produces two simultaneous tones which the telephone system decodes as the digit for that button. Tone dialing is still not available in some areas, but most of the United States is now covered.

After a connection is made, a small microphone in the telephone headset picks up your voice and amplifies it, sending it out on the two wires as an electrical signal. On the receiving end, another amplifier receives the

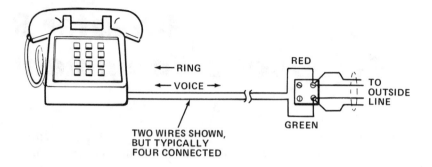

RING

VOICE →

RED

TO
OUTSIDE
LINE

GREEN

TWO WIRES SHOWN,
BUT TYPICALLY
FOUR CONNECTED

Figure 9-1.
Telephone Wiring

electrical signals, amplifies them and sends it to a small speaker in the headset.

Although this seems simple enough, a great deal of switching and computerized functions go on in the telephone lines once they leave your house. Telephone company services are increasingly sophisticated and offer such services as *call waiting, speed dialing,* and *call forwarding,* all controlled by telephone company computers. Some of these services, however, can be implemented in your own equipment. In addition, many other features can be implemented with your own telephone that the telephone companies do not currently offer.

Even though it takes only two wires to connect to a telephone, you may have a bundle of six wires inside your walls. There may be as many as three separate lines available in your home, if you have a fairly new house that has been *prewired* for phone service. Of course, the telephone company must hook up each new line outside your house and you'll pay for the extra service. However, there's no reason you shouldn't install your own phones, purchased from your own supplier with any options you'd like to have.

Telephone Options

Here are some of the options available on ordinary single-line phones today:

Pulse or Tone Dialing

Even though your telephone line may not be set up for tone dialing, you may be calling companies that decode tones for entering data. You've probably heard a recording more than once say "If you're calling about your bill, press "1" now. If you're calling about repair, press "2"…" When your phone has switchable pulse/tone dialing, you can switch to the tone mode and use tones for those services, even though you may not be able to dial out with tones.

Mute Button

A *mute button* cuts off the microphone in the headset so that you can hear the other party, but they cannot hear you. This is convenient for listening to the other party while you shout instructions to your children!

Touch Redial

A *touch redial* feature allows you to redial the last number dialed by pressing a button, usually the # key. This is handy for redialing a number that was busy the first time - and the second time - and the third time.

Adjustable Ringer Volume

Some phones provide a high/low switch for the telephone ring, or even a position to turn off the ring completely. Other models provide a continuously adjustable level. Bell sounds are a "natural bell" (original telephone bell sound) or several other types of bells or electronic signals.

Hold Button

A *hold button* allows you to keep a call connected, but to temporarily disconnect the phone audio - both your voice and the incoming voice. Usually there is a blinking indicator light indicating that there is a call on hold. If the call disconnects, the light will go off. Like the mute switch, the hold option is convenient for privacy.

Speakerphones

Some models of phones have a built-in *speakerphone*. This is a small amplifier that amplifies the incoming voice and plays it through a small speaker in the base of the phone so that several people near the phone can hear the conversation.

Adjustable Volume

Some phones have a built-in *amplifier* with adjustable volume. This is especially useful for the hard-of-hearing.

Prices

The options above are offered in various combinations on many different models and styles of phones - decorator phones, *slim phones*, *one-piece phones*, or standard desk phones. Prices range from about $9.95 (!) to $70.

Multiple-Line Phones

The garden variety phones described above are connected to a single line. However, two-line desk telephones are available for $60 to $100. A second line requires that your telephone company connects the line at a box outside of your home, but if your home is prewired with several sets of lines, a second line should pose no problem as far as your installation.

Two-line telephones have two buttons, one for each line. When the button for line 1 is pressed, you're connected to line 1. When the button for line 2 is pressed you're connected to line 2. There's also a *hold* button, which allows you to answer another line while not disconnecting from the first line. In fact, you can have two lines active at the same time, switching back and forth between them.

Telephone Dialers

Telephone dialers (Figure 9-2) are usually built into telephones, but there are some stand-alone units that connect in parallel with existing phones. We'll describe the built-in units here.

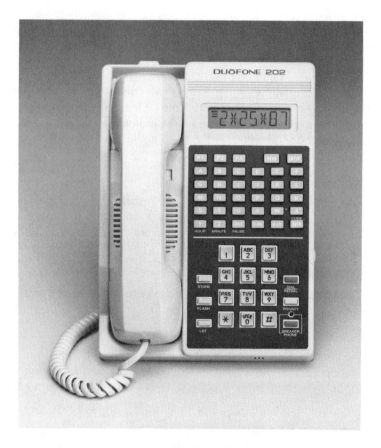

Figure 9-2.
Phone with Dialer
Courtesy of Radio Shack
A Division of Tandy Corporation

Telephone dialers will automatically *store* from ten to hundreds of numbers in a built-in internal memory. They will then dial that number automatically at the press of a single button, or after you've entered a two-digit code. Long-distance numbers or other complex numbers of up to 16 or 32 digits each may usually be stored and dialed.

One thing to watch for on auto-dialing phones is whether pulse or tone dialing or both is permitted. Some phones allow only pulse dialing, even on tone lines. Pulse dialing works fine, but is slower than tone dialing. Pulse dialing may take ten seconds for a seven digit number versus two seconds with tone dialing. Some pulse dialing phones operate at twice the normal rate - 20 pulses per second instead of 10 pulses per second.

Less expensive auto dialing phones simply dial the number at the press of a button. More expensive models have a *built-in display* - usually a liquid-crystal display - that allows you to see the number being dialed and to enter the number initially. Once you have a display on the phone other intelligent features keep creeping in! These phones also may display the current time, day of the week, elapsed time for calls, and other information.

The ultimate phone dialer with display is like a small computer. Not only does it display the current time and date, it will let you store numbers referenced by nicknames or the first few letters of a name. For example, suppose you had forgotten a phone number of a friend called JOE. You could enter J-O-E and the phone dialer would display the number it found corresponding to J-O-E in its memory! This type of intelligent phone also has other features, such as redialing a number dozens of times if it is busy.

One handy feature found on some auto-dialing phones is the ability to press a *single button* for emergency numbers. Some auto-dialing phones allow you to define three emergency numbers that are dialed by a single button press.

Cordless Phones

Cordless phones (Figure 9-3) are actually miniature short-range radio transmitters. If you have the proper radio set, you can actually hear one or both sides of a neighbor's conversation. However, the odds of someone listening in on a cordless phone are rather remote - they'd have to know when you were going to be making calls and also have the radio equipment.

Older cordless phones used two different frequencies for talking and listening, one at about 1.7 MHz, and one at 46 to 50 MHz. The frequency at 1.7 MHz was prone to more noise and static than the 46 to 50 MHz frequency. As a result, these phones are noisier than the newer models, which use two frequencies in the 46 to 50 MHz band, a much more noise-free radio band.

Cordless phones come in two parts, a base unit, which includes recharging capabilities for the batteries in the handset, and the handset, which you can carry around with you. Both the base unit and handset have a

Figure 9-3.
Cordless Phone
Courtesy of Radio Shack
A Division of Tandy Corporation

small telescoping antenna, which you can extend for greater range. Typical ranges for cordless phones are 100 to 200 feet. This range can be extended for farm, ranch, or business use by a beefed-up antenna system at the base unit.

Most models have a built-in *security code* provision which can be set to a secret code. This code is programmed into both units and prevents other people from accessing your telephone line.

Three factors should be checked when buying a cordless phone. All cordless phones have about the same transmitter power, but some have a greater range than others. If you will only be using the phone around your house, this is usually not a problem. The second factor is speech quality. There's a great variation in speech quality on cordless phones. Some seem muffled or distorted, while others have more clarity. The best way to test this is by actually trying the phone at a store, if that's possible. The last thing to check is that the phone has tone dialing if your telephone system allows it.

As you might suspect, some cordless phones have an auto-dialing feature built-in, typically allowing you to auto-dial 30 numbers or so.

Cordless phones range in price from about $80 to $200.

Telephone Answering Equipment

The telephone answering machine (Figure 9-4) has become very popular - it seems like everyone has one. Telephone answering machines are a convenience - you really never have to miss a phone call again, and the prices have continually come down. Telephone answering machines are priced at about $80 to $200.

All answering machines have the same basic capabilities. They have a cassette tape for your message and another cassette tape for recording calls (although newer models may have voice synthesis - see below). You can record a message, edit the message (listen to the message to see if it sounds all right), set the machine to answer calls, rewind and play back the calls, or set the machine to play back an announcement only. You can also *call monitor* a call - listen to the incoming call before answering it.

Here are some of the extra features found on different answering machines:

Figure 9-4.
Telephone Answering Machine
Courtesy of Radio Shack
A Division of Tandy Corporation

Remote Control

Remote control lets you call your answering machine, send a coded audio tone, and listen while the machine rewinds and plays back your calls for you. Once this is done, the machine is set up to record calls once again. This feature allows you to check your calls from just about anywhere in the world. An incredible feature on some models: when checking calls, the answering machine will answer on the first ring if there are calls recorded, but will answer after four rings if there are no calls. If the phone rings more than once, you can hang up and not be billed for the call, knowing you have no calls!

The *beeper* used for remote control may be a small hand-held unit about the size of pack of cigarettes, powered by a battery. Newer models are *beeperless*. They operate by pressing Touch-Tone buttons on the phone from which you're calling.

Tape Counter

Some answering machines have a *tape counter*. This is a convenience if you want to save some of the calls and don't want to rewind the tape completely, or if you want to check a certain call again.

Two-Line Answerer

Some answering machines have a *two-line* capability. One line is answered by a personal message, while the other is answered by a second *voice-synthesized announcement* to leave a message.

VOX and CPC

A *VOX* feature saves cassette tape by starting the tape only when the caller is speaking. *Calling party control* (CPC) also saves tape by stopping the recorder when the calling party hangs up. This eliminates a lot of wasted time wading through recorded material in which there is no message.

Time Stamp

In some answering machines, a built-in voice synthesizer *time stamps* each call - the current time is recorded before each message.

Answerer/Phone Combinations

Some phones have a built-in answering machine, or should we say, some answering machines have built-in phones?

Installing Your Own Phones

Most modern phone installations use a standard *modular* plug, shown in Figure 9-5. This plug has provision for four wires, colored black, red, green, and yellow. The color of the wires can be seen through the plug, which is transparent.

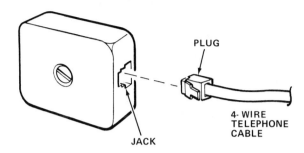

Figure 9-5.
Standard Modular Plug

Older style phone installations used a four-prong jack and plug shown in Figure 9-6. Adapters are available at Radio Shack to convert this older four-prong plug or jack to the newer modular plug, as shown in the figure.

We'll assume that you have a modular style jack, or have converted a four-prong plug to this type in the following discussion.

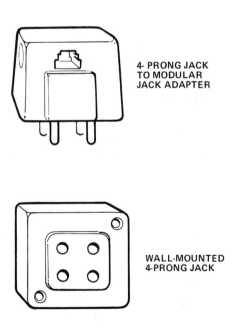

Figure 9-6.
Four-Prong Jack and Adapter

Telephone Cable

Telephone cable can be purchased at any Radio Shack store. This cable has four conductors and is color coded in black, red, green, and yellow. If you are adding new telephone lines for dozens of feet, get the *solid* conductor cable for easy screw-type connections. If you are making your own telephone cables with modular plugs at the end, get *modular* telephone cable, a 26-gauge cable with a gray outer jacket.

Crimping Tool

Modular *connectors* are available in a package of ten for about 30 cents each. These connectors fit the telephone cable described above. A *crimping tool* is also required to attach the modular connector to the cable. The crimping tool sells for about $9.

Connecting a plug to the cable is very easy and is shown in Figure 9-7. First, cut the cable squarely. Next, *strip* the cable using the built-in stripper on the crimping tool. This removes the outer jacket from the four conductors for a short distance. Next, insert the modular plug into the crimping tool. Now, insert the four conductors into the end of the plug, taking care to position the black and yellow ends as shown. Pressing down on the crimping tool makes the connection. The entire process takes less than a minute, even as little as 15 seconds once you have practice.

Adding More Plugs

It's easy to add another modular plug to an existing plug. Just get a duplex jack, shown in Figure 9-8. The duplex jack plugs into an existing modular plug and gives you two jacks. There are also three- or four-jack versions. The duplex jack can be used to connect telephone answering machines to an existing wall jack or to add an extension phone.

Extending a Telephone Cord

To extend a telephone cord, buy an *in-line coupler*, shown in Figure 9-9 and a 12- or 25-foot *modular to modular* cable. You can even make your own cables of virtually any length using the crimping tool, telephone cable, and modular connectors described above.

Adding a New Wall Jack

Follow these steps to add a new wall jack for a telephone extension. First, examine the existing phone connections to see which color-coded cables are used. Usually red and green are used for the first line and yellow and black are used for a second line for a four-wire cable. Six-wire cables are also common and use three pairs - blue and blue/white (line 1), orange and orange/white (line 2), and green and green/white (line 3).

Next, find the dummy wall cover plate in the room in which you want to add the wall jack. If there is *no* wall cover plate, you'll have to run a four-wire

STEP 1: CUT CABLE SQUARELY

SQUARE
END

STEP 2: STRIP CABLE

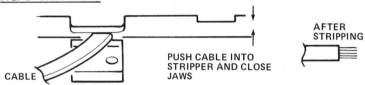

AFTER
STRIPPING

PUSH CABLE INTO
STRIPPER AND CLOSE
JAWS

CABLE

STEP 3: INSERT MODULAR PLUG INTO TOOL

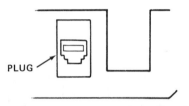

PLUG

STEP 4: INSERT CONDUCTORS INTO PLUG AND CLOSE

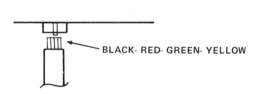

BLACK- RED- GREEN- YELLOW

AFTER ASSEMBLY:

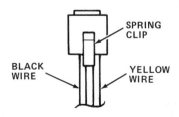

SPRING
CLIP

BLACK
WIRE

YELLOW
WIRE

Figure 9-7.
Connecting Modular Plugs

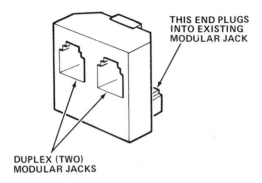

THIS END PLUGS
INTO EXISTING
MODULAR JACK

DUPLEX (TWO)
MODULAR JACKS

Figure 9-8.
Duplex Jack

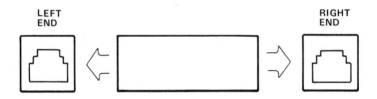

LEFT
END

RIGHT
END

Figure 9-9.
In-Line Coupler

telephone cable along the baseboard from an existing phone line, securing
the line with insulated staples. In this case, connect the wires as shown in
Figure 9-10.

If you do have a dummy wall cover plate, remove it and install a wall
phone plate with a modular plug. Connect the same color-coded wires as in
the existing phone to the cover plate terminals, as shown in Figure 9-11.
Before replacing the cover plate, plug in a phone to see that it works. Now
secure the jack to the wall - usually the two screw holes match standard
junction box spacing.

If you've run a cable, use a surface mount modular jack, as shown in
Figure 9-12. Connect the same color-coded wires as in the existing phone to
the terminals of the surface mount jack. Plug in a phone and test the circuit
before replacing the jack. Now secure the jack to the wall or baseboard.

OLD 42A STYLE

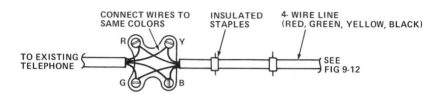

OLD 4-PRONG STYLE

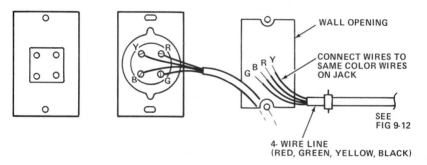

NEW MODULAR STYLE

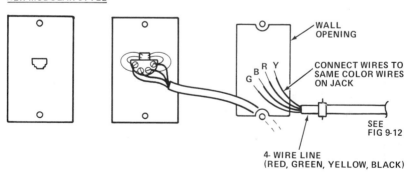

Figure 9-10.
Extending an Existing Telephone Line

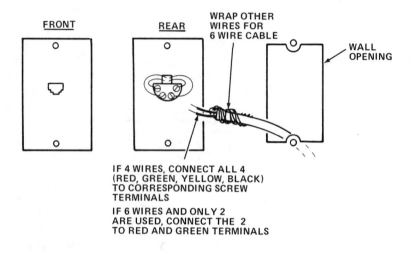

Figure 9-11.
Connecting to a Wall Phone Plate

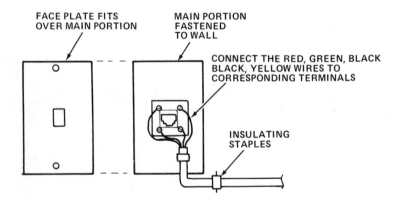

Figure 9-12.
Connecting to a Surface Mount Jack

Chapter 10
Car Telephones and Paging

Mobile telephones have been around for over 40 years. However, *cellular car phones* are a new type of mobile telephone which offer high quality phone service. Paging is another type of radio service which can be used over a wide area to let you know that there's a message or call waiting for you.

Cellular Car Phones

General

All mobile telephones are low-powered radio transmitters and radio receivers. Cellular phones are no exception. Their power is about that of a CB radio - three watts. Normally, three watts would provide telephone service which could not cover a wide area. In the Los Angeles area, for example, the city and suburbs extend over a 70-mile range, which a three watt transmitter could not reliably cover.

However, cellular phone systems divide a metropolitan area into *cells*, hexagonally-shaped small segments as shown in Figure 10-1. (Actually the cells are circles, but hexagonally-shaped figures can be more easily "tiled" together on maps.)

As you use the car phone while driving, you move from cell to cell. Cells are located a few miles apart, so you may move across several cells in a 10-minute time. Each time you enter a new cell, you come under control of a new master transmitter/receiver that receives your signal and broadcasts the signal to which you are connected. This "handing off" to a new transmitter/receiver occurs automatically, under computer control - you are usually not even aware that it is happening.

Since each cell is only a few miles across, the result is *local* communication in which you get clear, high-quality conversation.

Your call is routed automatically to two channels (one for transmitting and one for receiving). Selection is made from the hundreds of channels for that cell. Since each cell is not that large in area, there are usually more than enough channels to go around.

Each cell, in turn, is connected to a centrally located MTSO, or *mobile telephone switching office*. The MTSO ties together all of the cell transmitter/

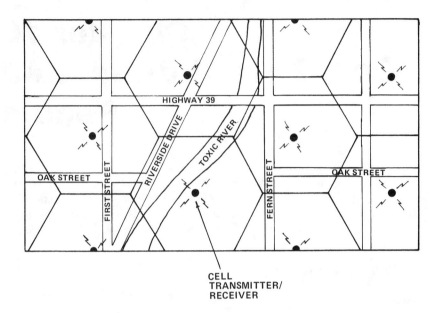

CELL
TRANSMITTER/
RECEIVER

Figure 10-1.
Cellular Phone Cells

receivers and connects them into the standard phone system. From there, your call is routed over telephone "land lines" to connect you to a party on a conventional telephone. (Of course, you can also talk to another cellular car phone.)

While driving in your car, you can be reached by a calling party the same way you are reached by a conventional telephone. The MTSO locates your preassigned cellular phone number and rings the car phone just as it would ring your phone at home.

Coverage

Cellular car phone service is available in metropolitan areas, such as Los Angeles, Las Vegas, Chicago, Phoenix, Dayton, Little Rock, New York, and many other cities. In the Los Angeles area complete coverage is provided across a 100 mile diameter area. Other cities are comparable. Coverage is not generally provided when you drive *out* of the metropolitan or suburban area, however. In driving from Los Angeles to Phoenix, for example, you will not be able to get car phone service from the outskirts of Los Angeles to the outskirts of Phoenix.

However, the good news is that the telephone company with which you have service usually also has service in other metropolitan areas. Even if it has no offices in that area, it coordinates service with other *carriers* so that you will be able to use *their* cellular services. This *roaming* capability is usually automatic - you can simply drive into an area and start using the service, as you would in your local area (typical roaming charges are $.15 per minute).

Within your local area, coverage is usually continuous, without breaks. For example, you could drive the length and breadth of the Los Angeles area and carry on a conversation all the way. However, there may be occasional pockets where reception is bad. This usually lasts only for a few seconds, but occasionally may cause a disconnect. Terrain is usually the problem in these cases - high mountains or hills may temporarily block the signal.

Security

Cellular car phones provide good, but not perfect, security. Older car phones were easy for a radio enthusiast to monitor. Conversations can be heard on a single frequency with an inexpensive receiver. Both sides of the conversation can be heard. However, cellular car phones use frequencies in the 800 megaHertz region, beyond the range of much radio equipment. Furthermore, since individual cells are used and there is no single powerful transmitter/receiver, it's difficult for a person to monitor the individual cell transmitters. This is especially true because each cell transmitter/receiver may switch to another frequency from the preceding cell. Each call uses two different frequencies, one for the calling party and one for the receiving party. While a motivated person can use radio equipment to monitor some cellular phone transmissions, about the best he can do is to hear snatches of short conversation from one party and he could not easily find a specific caller.

Typical Cellular Installations

Cellular phones actually come in two types. A *portable cellular* phone may be carried around in a small pack and used anywhere. The power of such a unit is generally about 3/5th of a watt, about one-fifth of the three watts of a car phone; the units, therefore, are a little more susceptible to poor signal quality.

Cellular *car phones* are permanent installations in cars or trucks. There's a small handset unit in the driver's area, a transmitter/receiver unit, generally mounted in the trunk, and a cellular phone antenna, as shown in Figure 10-2.

The phone unit is somewhat the same as a small desk telephone, as shown in Figure 10-3. There are push buttons to dial the number, a SEND button to connect to the line, an END button to disconnect, and buttons to store and retrieve numbers to dial, just like an automatic telephone dialer.

Typical cellular phones allow you to store up to 32 or 40 commonly used telephone numbers. The numbers may then be dialed by pushing one or two buttons.

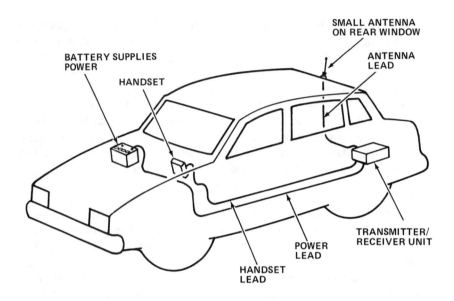

BATTERY SUPPLIES POWER

HANDSET

SMALL ANTENNA ON REAR WINDOW

ANTENNA LEAD

TRANSMITTER/ RECEIVER UNIT

POWER LEAD

HANDSET LEAD

**Figure 10-2.
Cellular Car Phone Installation**

If you're out of the vehicle when a call comes in, some cellular car phones will alert you to this fact by sounding the horn or flashing the headlights.

Cellular car phone antennas are usually less than a foot in length. Although they can be mounted on a car's roof, a more common mounting is on the rear window. Although the rear window mounting *looks* like the antenna goes through the window via a small hole, in fact there is no hole in the window. There's a small plate on either side that couples the inside and outside portions of the antenna together through the glass.

Cost

Car phone prices have dropped steadily. Cellular car phones now cost from about $900 to $2000, including installation. There's usually a one time connection fee (typically $50) and a fixed monthly fee (about $45). Two hours of phone use per month for local calling might cost about $88. However, 20 hours of phone use per month would be on the order of $480, although volume rates are available. Installation and connection can usually be done in one or two days. If you spend even one or two hours a day in your car, cellular car phones will allow you to put that time to good use.

Paging Devices

Cellular car phones allow a two-way conversation from your car. *Paging devices*, however, are a radio device which do not allow a conversa-

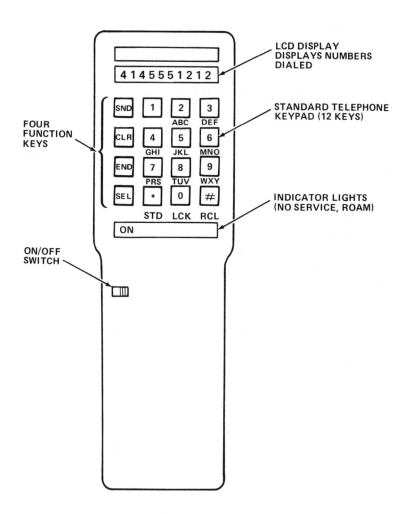

Figure 10-3.
Cellular Car Phone Keypad

tion at all - they simply beep to signal you. The advantage of pagers, though, is that they can be used over a wide area and can signal you just about wherever you are.

Paging Services and Pagers

Pagers operate in conjunction with a *paging service*. You pay a monthly fee for such a service, typically $8 to $20 per month. The paging service has a high-powered radio transmitter which can cover an area of thousands of square miles - the size of a large metropolitan city.

To use a paging service, you carry a *beeper* on your belt or in your pocket. Another term for the beeper is a *personal pocket pager*. When someone wants to page you, they call your paging number on the telephone. This call automatically triggers a radio signal that is received by your personal pocket pager. The pager then emits a beep to let you know that someone is trying to get in touch with you. You can then call in your office to find the phone number of the caller, or call later to get the message left by the caller.

Some personal pocket pages have a *silent paging mode*. The pager records the fact that a page has been received, and the next time you check it, you can see that there is a pending page. This mode is handy for occasions during which you do not want to be disturbed with an audible signal.

Pagers may also have several different codes. A different beep sequence is sounded for each code. Each code has an additional charge. (You might, for instance, want to use one code for a personal call page and a second for a business call page.)

Another type of pager is a *digital* pager. In this type of pager, the caller leaves his phone number, which is then displayed as digits on the pocket pager display. This type of pager allows you to call the pager directly instead of first having to go through a central number. Digital pagers are about twice as expensive as the audio type.

Figure 10-4 shows a typical pocket pager and its functions. This particular pager costs $99, but other pagers range in prices from $90 to $180.

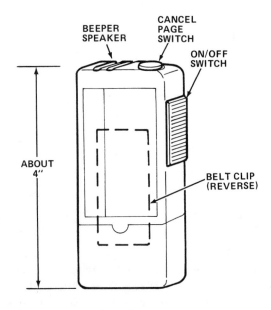

Figure 10-4.
Pocket Pager

Local Wireless Pagers

Don't want to pay a monthly service charge for a paging service? You can install your own paging service! A *personal paging service* (Figure 10-5)

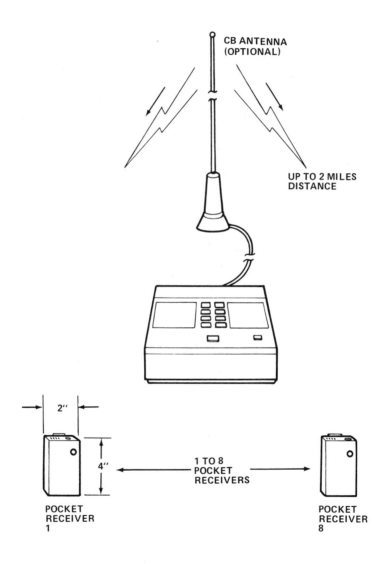

Figure 10-5.
Local Wireless Pager

uses a transmitter that is located in your home, farm, office, or manufacturing facility. You can page any one of eight people that are carrying pocket receivers by pressing one of eight buttons. The local transmitter sends out a signal that is received by the person anywhere within about a two-mile radius. (Actually the distance may be up to five miles radius or so with a suitable outdoor antenna.)

The pocket paging receiver is very similar to the one carried when you subscribe to a paging service. The transmitter and receivers can be coded to a special code that prevents triggering of the pocket receiver by a false signal.

This type of pager is handy for contacting local personnel in your own business facility when they are on the grounds or not far off the facility.

Typical paging systems of this type cost about $100 for the transmitter and one receiver and about $30 for additional receivers. An external antenna costs about $25 additional.

Chapter 11
Auto Electronics

Auto electronics can be divided roughly into audio components such as tape players and radios, radar detectors, auto alarms, and CB radios. Many of the same electronics devices that can be found in a home are also present in a car or truck, although scaled down in size. We'll describe what's offered for cars, trucks, or mobile homes in this chapter.

Audio Components

Audio components duplicate the components you'd find in your home for audio entertainment. There are AM/FM/cassette players, equalizers, amplifiers, power boosters, and speakers. A block diagram of a typical automobile audio system is shown in Figure 11-1.

Antennas

A car antenna is a necessity if you want to receive AM and FM radio. Chances are your car already has one installed, or possibly has a "knock-out" hole for an antenna. The type and length of the antenna for AM is not as critical as for FM radio. Just about any length of antenna will do for AM, which is why you'll see people with bent coat hangers serving as car antennas! For FM, however, a good antenna will help pull in the stations and prevent fading as the car changes direction. No special antenna is needed for FM stereo - any antenna that also receives monaural (one channel) FM will also work for FM stereo.

Most antennas come with a cable connected to the antenna on one end and with a *Motorola* type plug, as shown in Figure 11-2. The Motorola plug matches the female Motorola jack found on most car radios.

Am/FM/Stereo Radios

Most car radios now receive AM stations, FM stations, and include a built-in cassette tape player. Not too many years ago, eight-track stereo was popular, and magnetic tape players were designed for eight-track cartridges, but cassettes are now far and above the most popular type of tape.

AM/FM/Stereo radios start at about $70 and go up to $250 or more. Here are some of the options to look for:

Power

One of the biggest factors affecting the price is *power output*. Only about two or three watts of output power are required for quiet road conditions. However, some people like their music loud, and you'll find AM/FM/Stereo radios that deliver from about eight to 20 watts per channel (40 watts total). One reason for buying a more powerful unit is that there is less distortion when the music *peaks* on a loud passage. Also, listening conditions are usually not as quiet as they are at home - often there's a lot of road noise that can be covered up with soothing (or not so soothing) music.

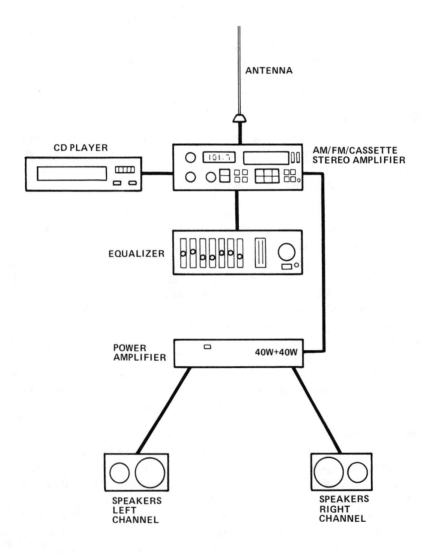

Figure 11-1.
Auto Audio System

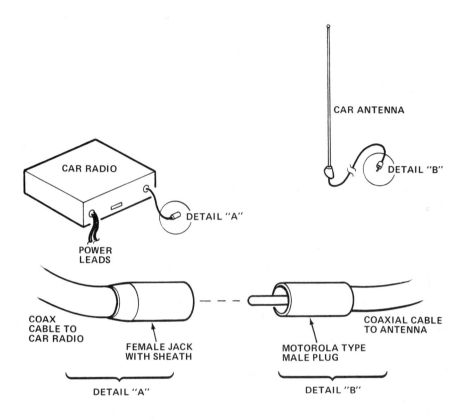

Figure 11-2.
Antenna Jack for Car Radio

Slide-Rule Versus Digital Tuning

The least expensive AM/FM/Stereos have *slide-rule* tuning for the radio. In this type of tuning, a dial pointer slides across a dial. Most radios, however, have *digital* tuning. In digital tuning, the frequency of the station selected is displayed on (usually) an LCD (liquid crystal display).

Digital tuning also provides a *memory* - the frequency of stations that you listen to frequently can be stored in the memory of the unit, and then station selection can be made by pressing a single button. This is a handy feature for driving, as it requires less time with your eyes off the road. Typically, about six AM station and six FM station frequencies can be stored in memory.

Tuning can also be done manually, with a dial, or you can set the tuner to go into a *seek* mode to select the next station or to scan the available stations.

Digital tuners skip in 0.01 megaHertz increments on AM and on 0.2 megaHertz increments on FM. The reason for this is that stations are pre-assigned in these frequency increments. Most United States stations are exactly on frequency, as well, although occasionally you'll find a station slightly off, resulting in distortion of the audio. (With most tuners there's no way to tune slightly off frequency to receive such a station - such is progress!)

Digital tuners are all *quartz tuners,* which simply means that the frequencies are constructed from electronic crystals. Quartz tuning is a very accurate form of tuning.

Cassette Player Features

More expensive AM/FM/Stereos have *Dolby noise reduction.* Dolby noise reduction is electronic circuitry that reduces tape hiss at low volume. It improves tape performance considerably.

Faders and Balance

Stereo in a car is a little more complicated than stereo at home. There is often more of an imbalance in the sound from one channel over the sound from another channel. Also, there may be front and rear speakers installed in the car. There is usually a *balance control* to adjust the levels of both the right and left channel. In addition there is often a *fader control* to balance the four speakers often used in cars.

Tuner, Cassette, and Amplifier Specifications

It's often hard to find specifications about the frequency response of the tuner, cassette player, and amplifier for automotive audio equipment. Suffice it to say that the same specs apply as for home audio equipment. It may pay you to read Chapters 1 and 2 and apply the hints about frequency response, distortion, and cassette specs found there to evaluate audio equipment for your car.

Separate Cassette Players and CD Players

If you already have a good AM/FM radio but want cassette capability, you can buy an individual cassette player to mount under your dash. Prices range from about $30 to $100. One possible problem, however, is switching the car's speakers between the radio and cassette. You'll probably end up buying a speaker switch as well, and wiring can be complicated for four speakers.

Also available are small CD (compact disk) players, which can be dash-mounted as well.

Speakers

AM/FM *speakers* are similar to home speakers in an audio system. Speakers may consist of a single speaker for each of the two channels of sound, or there may be two or more speakers per channel. The two sets of speakers should be the same type - don't buy two different speakers for the right and left channel, as the channels will not match in sound quality.

Better speakers will have a *woofer, midrange,* and *tweeter* speaker. Each of these handles a different range of frequencies. All speaker sets that contain more than one speaker will have a built-in *crossover* network. The crossover network separates the sound into two or three frequency ranges, one for each speaker.

The speakers should be rated at least as high as your AM/FM/Stereo power output, or even   higher. For example, if your radio is rated at 20 watts per channel, each pair of speakers should be rated at 20 watts or greater.

Generally, you get what you pay for in speakers. The more expensive the speakers, the better the sound quality will be. Like home audio, speakers are the weakest link in an automotive sound system. If you're going to spend money anywhere, speakers are a good place to do so. Car speakers range in price from about $15 per pair to $100 per pair.

AM/FM/Stereo Installation

There are three external connections for AM/FM/Stereos. The first connection is to the antenna as described above.

The second connection is for power. All car radios and amplifiers require 12 volts dc, the voltage of most car batteries. Most radios require that the battery connection is a *negative ground,* although some will work with a *positive ground.* To find out how your car's battery is connected, look in your owner's manual, or simply open the hood to the engine and check the battery connections. The two battery terminals are labeled "NEG" or "-" and "POS" or "+". If the NEG/- post is connected by a heavy battery cable to the car frame, as shown in Figure 11-3, your car has a negative ground. If the POS/+ post is connected by a heavy battery cable to the frame, your car has a positive ground.

If your car has a positive ground, *voltage inverters* are available that will change the positive ground to a negative ground so that you can install a negative ground radio.

The power cable is usually *fused* and labelled. Your car may have existing leads for a radio behind the cover plate on the dash. If you're replacing an existing radio, the power leads will also be obvious.

Power connections vary with the car's manufacturer and conversion kits are available from Radio Shack to adapt existing car wiring to the equipment in use.

The third connection is to speakers. Chances are the radio has a four-wire cable for speakers. Two of the wires go to the speakers for the right channel and two of the wires go to the speakers for the left channel. None of the wires should be grounded, but should go directly to the speaker terminals. You may have to route the wires under floor carpeting and through

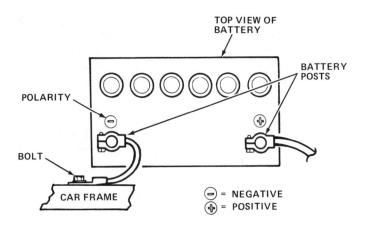

Figure 11-3.
Battery and Ground Connections

panels to reach the speakers, but the wires carry low-voltage currents and can be routed anywhere without problems. Just make certain that the wires do not continually rub against metal, as the insulation may wear through and short out the connections.

Speaker cabling is again available from Radio Shack.

Radar Detectors

Radar detectors (see Figure 11-4) are used to detect police speed radar. Police radar guns send out a burst of microwave energy. Microwaves are extremely high-frequency radio waves in the same bands used for line-of-sight communications between spacecraft. When the radar wave hits any object it is reflected back to a receiver in the radar gun. If an object is not moving, there is no shift in the frequency of the reflected radar wave. However, if an object is moving the reflected radar wave's frequency is changed slightly in a *doppler shift*. The effect is similar to a train whistle at a railway station. As the train approaches, the whistle sounds high pitched, but as the train passes you and moves away, the whistle seems to lower in pitch.

Radar detectors are really microwave receivers that tune the X and K microwave bands looking for radar signals. If any are detected, the detector provides a warning tone and/or light. The range varies with the type of radar being used and the terrain outside the car, but radar detectors may work even several miles away from the radar site. The warning tone generally beeps faster and faster as the the source of the radar is reached.

Depending upon the options, radar detectors range in price from about $100 to $200. Here are some of the things to look for in a radar detector:

Figure 11-4.
Typical Radar Detector

Sensitivity

The more *sensitive* a radar detector is, the greater the distance over which it will be effective. Sensitivity is measured in dBm/cm^2. This is a measure of the small microwave energy detected - the important thing to note is that the smaller this value is, the more sensitive the detector (disregard the minus sign). Sensitivity is usually less for the X-band. Typical sensitivity is -110 dBm/cm^2 for X-band and -102dBm/cm^2 for K-band.

Dual-conversion superheterodyne detectors are less susceptible to false signals (the "dual-conversion" part) and more sensitive to weaker signals (the "superheterodyne" part).

Controls

Most radar detectors have a high/low volume control. Better sets have an adjustable volume control so that you can set the warning tone to any sound level.

All sets have a sensitivity control. This control selects either city or highway sensitivity levels. In cities there are many sources of microwaves other than police radar - motion detectors, telephone links, commercial microwave, and others. The "city" sensitivity is not as sensitive as the highway sensitivity and reduces these spurious signals.

Size

Small size is an obvious advantage with radar detectors. Most detectors either clip to a sun visor, attach with suction cups to the windshield, or mount on the top of the dash. Some radar detectors can be concealed under the dash. A typical size is 3/4 inch by 2 inches by 4 inches.

Installation

Like car radios, radar detectors usually require a 12 volt negative ground system, although some will work with either negative or positive grounds. Many plug into the lighter socket on the dash, but a more permanent connection to the radio power supply leads can be made. An advantage to lighter socket power is that the unit can be rapidly removed and placed out of sight of highway police who may be biased against such devices.

Auto Alarms

Auto alarms will detect if your car is broken into, jacked up, towed, or been tampered with. Auto alarms are similar to the types of detection done in home security systems (see Chapter 8) - there are plunger type switches (*pin switches*) that connect to doors, hood, and trunk lid and are activated by opening or a vibration detector switch that activates if the car is jiggled. There are also small motion detectors that can be mounted inside the car and respond to any movement once set.

When the alarm is activated, some auto alarms continuously honk the car's horn or sound a loud siren mounted under the hood. Other alarms, amazingly enough, trigger a radio transmitter. The radio transmitter is attached to your car's antenna and sends out a radio signal. The radio signal in turn is detected by a pocket radio receiver that you carry. This means that you can detect break-ins of your car up to several miles away and take appropriate action.

Other auto alarm systems can be activated by a *panic button* mounted on a key chain remote control. This type of control can attract attention when you are confronted with a threatening situation in a parking lot or other location.

Figure 11-5 shows a typical installation of a *keyless* auto alarm that arms after the last door is shut. The alarm has a extra-loud siren mounted in the engine compartment.

CB Radios

CB radios let you talk to other cars, call for emergency help, and communicate with home-based stations. CB radios use a radio band called the *Citizens Band*, which is just below the VHF television band in frequency. Using this band, you can communicate dozens of miles locally.

All CB radios are regulated by law to low power - five watts. (To give you an idea how powerful this is, AM radio stations generally use 1000 to 10,000 watts of power.) However, this power is sufficient to provide reliable communications over a five or ten mile range.

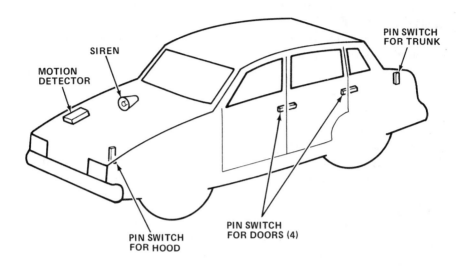

Figure 11-5.
Auto Alarm Installation

CB radios are excellent in road use. Truckers use CB radios to communicate with other truckers about traffic delays, road and weather conditions, police activity (highway patrolmen are called "smokies" in CB jargon), and just for company on the road. If you have a CB radio you can listen in on these reports even if you're too shy to use the radio for transmissions yourself. However, in an emergency situation, you can use the CB to call for help. Since many truckers and even volunteer groups such as REACT monitor emergency calling frequencies, a CB radio is a good way to summon help.

Operating a CB Radio

You don't need to be a radio genius to operate a CB radio. Figure 11-6 shows a simple CB radio setup. There's a microphone with a push-to talk switch - you just hold down the switch and talk into the microphone.

The microphone attaches to the CB radio proper. On the simplest CB radio there are only three controls. A *channel selector* selects one of 40 channels to monitor or on which to talk. These channels are numbered 1 through 40. Channel 9 is the emergency channel, which is monitored by many drivers or fixed stations. Channel 19 is commonly used by truck drivers.

After selecting the channel, the *squelch* control is adjusted. The squelch control turns on the receiver audio circuits only when there is a signal coming through on a channel. If no one is on the channel, there is a great deal of static hiss, and the squelch eliminates this tiresome noise when no one is talking on the channel.

The *volume* control acts like a volume control on any radio. It adjusts the sound level on the receiver.

To communicate over the radio, select the channel, adjust the squelch and volume controls, press the microphone push-to-talk switch and talk. To listen, let up on the microphone push-to-talk switch. And that's all there is to it!

CB Radio Buying Guide

CB radios are priced from about $60 to $200. Here are some of the extras that can be added to the basic radio:

Antennas

CB radios normally use another antenna from your normal car radio antenna. This antenna may be mounted on the trunk lid or top of the car roof. Special CB antennas with magnetic mounts may be used to temporarily install a CB antenna on the car roof. As a matter of fact, at least one system is a "road emergency system" which can be installed from a traveling case in seconds - plug the power cord into the cigarette lighter, place the magnetic mount antenna on the roof, select the channel, and call for help!

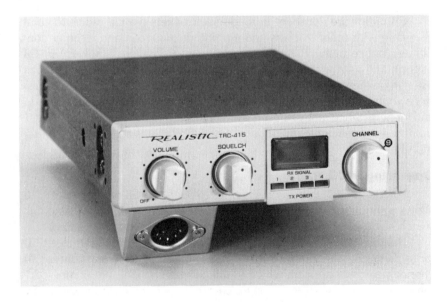

Figure 11-6.
Simple CB Radio
Courtesy of Radio Shack
A Division of Tandy Corporation

One-Touch Channel Selection

A *one-touch channel selection* allows you to switch to emergency channel 9 or to other commonly used channels with a single button push. The channels can be *programmed* in the unit.

RF Gain Control

Some CB sets have an *RF gain* control. This allows you to reduce the amplification of the receiver and prevent overload from strong signals which would otherwise sound distorted or not be understandable.

Noise Circuitry

Some CB sets have an *automatic noise limiter* (ANL) to reduce radio noise such as lightning crashes and static. Another type of noise reduction is a *noise blanker*, which reduces ignition noise.

Public Address System

A *public address* amplifier (PA) is found in some CB models. By throwing a switch, you can convert your CB set and microphone into a public address system, provided you have a PA remote speaker, typically mounted under the hood.

Signal/Power Meters

Signal and power meters provide a visual indication of how strong an incoming signal is or how strong your own signal output is. Usually, these signals are displayed on the front panel of the CB set with LED (light emitting diode) indicator lamps, although occasionally there will be an actual meter with dial.

Single Sideband CBs

The most powerful CB set is called a *single sideband* (SSB) set. Single sideband reduces part of the transmitted signal to pack more effective power into one *sideband*. The resulting signal is able to *punch through* static and noise more easily and can be heard over greater distances. SSB CB sets generally cost more money because they contain more circuitry, They are also somewhat more difficult to operate, as signals require a *clarifier* control to tune in exactly - without a precise tuning, the transmitting station resembles Donald Duck's speech.

The majority of CB users do not use SSB, but it is becoming more popular.

CB Licenses

CB radio is designed to be a "people's radio" to be used by anyone without knowledge of radio theory. Even a license is (no longer) required.

CB Radio Installation

Installation of a CB radio is very similar to installation of an AM/FM/Stereo unit. However, the CB radio requires mounting under the dash or other convenient place, rather than in the preassigned radio location found in most cars and trucks. Also, most CB sets have a self-contained speaker and require no external speakers, although you might want to use one for better sound quality.

The hardest part of CB installation is mounting of the antenna, if a permanent antenna is required. This may require drilling a hole in a trunk lid or rooftop, and routing the antenna wire. An alternative to this is replacing your AM/FM antenna with a combination CB/AM/FM unit. Although not quite as effective an antenna for CB, the combination antenna eliminates what some people would regard as an unsightly second antenna.

Chapter 12
Other High-Tech
Goodies

In addition to audio and video equipment, security systems, telephones, and auto electronics, there are many other high-tech electronic devices. *Short-wave radios* and *scanners* are used to monitor commercial and other broadcasting. *Walkie-talkies* can be used for short-range, portable communications. *Calculators* range from simple adding machines to specialized portable computers.

In the hobby area, *electronic keyboards* can produce an amazing variety of musical sounds at inexpensive prices. There's also a host of *electronic toys*, which grow more interesting and sophisticated every year. Also in the category of electronic toys are *electronics kits* which provide educational training in radio, computer science, and electronics.

Finally, there's a wide range of *electronic test equipment* which can be used to test and diagnose problems in home electrical and electronic devices.

Radios and Scanners

In addition to the normal AM and FM bands that broadcast music and entertainment, there's a huge spectrum of radio frequencies used for other purposes.

Most countries have *short-wave* broadcasting of news, entertainment, sports, and cultural programs. These short wave broadcasts are easily received in the United States and worldwide with *short-wave receivers* (see Figure 12-1). Using such a receiver in the evening, for example, it's easy to hear English-language broadcasts from Radio Moscow, the BBC (Great Britain), Radio Canada, and Radio Japan. Although reception is not as crystal clear as local stations, short-wave broadcasts are dependable and of good quality. Good short-wave receivers can be purchased for under $100. Make certain that such a receiver covers at least the 49-meter and 41-meter short wave bands.

For the more serious radio listener, there are commercial radio broadcasts to ships at sea, international airplane communications, amateur radio communications (ham radio), and military communications. There are even pirate radio stations and coded messages to be heard. Most of these communications can be tuned in with short-wave receivers as well, although certain

Figure 12-1.
Shortwave Receiver
Courtesy of Radio Shack
A Division of Tandy Corporation

types of communication require a *communications receiver* with a built-in circuit for receiving morse code called a *beat-frequency oscillator* or *bfo*. Communications receivers generally cost from $250 to $500.

For receiving occasional voice short-wave broadcasts only, most short-wave receivers will do, even the inexpensive variety. If you do serious short-wave listening, a *digital* receiver has amenities such as programmable tuning and station selection. If you also want to monitor amateur radio code, commercial stations, and other communications, buy a *communications receiver* with a *bfo*. The latter can be can be purchased for as little as $200.

Local communications - police and fire, ambulance service, taxi cab and truck dispatchers, amateur radio operators, aircraft communications, and mobile telephone - use a type of radio communications called *VHF* and *UHF FM broadcasting*. The frequencies used are in the same range as FM or tv broadcasting. These types of communications are local only and are limited by line-of-sight distances.

Scanners (Figure 12-2) are radios that allow you to tune in these frequencies and *monitor* communications, whether it's police and fire services or aircraft controllers at local airports. Two general types of scanners are used, *crystal controlled* and *synthesized scanners*. The former are limited in frequency coverage - only a few channels can be scanned, and a *crystal* must be installed for each channel. The more expensive *synthesized scanners* do not use crystals and can monitor hundreds of separate channels.

Even if you monitor only a few channels, the *synthesized* type of scanner is a great convenience. Furthermore, the price difference between crystal-controlled scanners and synthesized scanners is not that great.

Weather radios monitor only weather channels in the same frequency range. Weather channels are local radio stations run by the Weather Service (NOAA) that broadcast continuous accurate weather reporting. Some weather radios include the capability of sounding a warning tone when NOAA broadcasts weather emergency information, such as impending tornado activity.

For more on these products, refer to *Shortwave Listeners Guide*, a Radio Shack book by the same author.

Figure 12-2.
Scanner
Courtesy of Radio Shack
A Division of Tandy Corporation

Walkie-Talkies

Far from being children's toys, *walkie-talkies* are useful communications devices. They can be used, for example, for reliable two-way communications on camping trips, for communications on building projects ("How's the tv picture with the antenna pointed *this* way?"), and for other practical two-way communications.

There are two basic types of walkie-talkies, those that use the 49-MHz band and those that use the 28-MHz Citizen's Band (CB). The 49-MHz band type is limited to 1/10-watt output, which is sufficient for communications over 1/4 to 1/2 mile or more, depending upon the terrain. This band *is* used for children's walkie-talkies, but the same devices can be used for serious communication if the range is not too great.

The 28-MHz CB walkie-talkies can use up to 5 watts of input power. Input power for these devices range from 1/10 watt (100 mW) to 5 watts. The higher the input power, the greater the distance that may be covered. However, doubling the power *does not* double the distance - increasing the power 10 times will double the distance. A 5-watt CB walkie talkie is good for communications of about one or two miles or more, depending upon the terrain.

CB walkie-talkies (Figure 12-3) are more expensive than the 49-MHz version. Most require crystals, but can cover several channels with multiple crystals. This is important because the CB channels are extremely crowded and you may experience interference from another station.

For children's toys, the 49-MHz version of the walkie-talkie is a good buy. For serious, reliable adult communications of up to two miles, get a CB band version with three- to five-watt power input, and at least two channels.

Calculators

Calculators range from credit-card sized to printing calculators and are portable enough so they can be used in the grocery store, in the field, or at school. Calculators are different from small *computers* in that calculators:

- are very portable, usually vest pocket size

- are self-contained, requiring no additional equipment

- provide only a one-line display, in most cases

- have less programming ability than a computer, or none at all

- are much less expensive than a small computer

Figure 12-3.
Walkie-Talkie
Courtesy of Radio Shack
A Division of Tandy Corporation

There are four general types of calculators.

A *basic* calculator (see Figure 12-4) performs addition, subtraction, multiplication, and division and has the ability to save temporary results in memory. It does calculations in mixed numbers as well. For example, you could add 123.45 and 34.7 and divide the result by 34.66 to get an answer of 4.562896711, displayed on the *LCD* (*liquid crystal display*) of the calculator. In addition to these basic arithmetic operations, the basic calculator may also do automatic square roots, percentages, and sign change (sign change allows you to change a result of 123.87 to -123.87 with a single key instead of having to reenter the number). Basic calculators are perfect if you're using them for adding up prices in a grocery, balancing the checkbook, or perform the simple calculations that most of us do day by day.

Simple calculators range in size from a thin "credit card" calculator 1/8 inch thick, to a larger desktop calculator that can be placed on a desk in a more or less permanent position. Prices range from under $10 to $50.

A *printing calculator* performs the basic operations of the calculator described above, but also prints out the entries and results on narrow paper, typically 2-1/4 inches wide. The advantage of the printing calculator is that there is a permanent record that may be annotated and filed, or given as a receipt. Prices of printing calculators range from $40 to $100.

Figure 12-4.
Scientific Calculator

Courtesy of Radio Shack
A Division of Tandy Corporation

Business calculators provide the basic functions above, but also have several dozen business related functions. You can, for example, figure the monthly payments for a $10,000 loan at 12.75% per year over 5 years at the press of a few buttons. Furthermore, the result would be as accurate, or even more so, than the same problem run on a large computer system. Typical business calculators provide easy solutions to problems such as interest, bond rate conversions, and many others. The cost of this type of calculator is on the order of $25 to $90.

Scientific calculators provide the basic functions, but also have dozens of scientific functions built in. This type of calculator can find the sine value of an angle in degrees or radians, provide a random number, and calculate the standard deviation for statistical applications - functions that a basic calculator could not perform, or could perform only with a great deal of trouble. Scientific calculators also provide memory for storage of inter-mediate results. Some scientific calculators also are *programmable* - the steps to perform a calculation can be stored in memory and then executed in sequence to perform the calculation.

There are two basic types of scientific calculators. One type, the Hewlett-Packard type, uses a type of mathematical notation called *reverse Polish notation*. Adding two numbers, for example, is done by entering 123, followed by 567, followed by an addition sign.

The second type of calculator uses the normal *infix* notation. To add 123 and 567, you'd enter 123, an addition sign, and 567.

Scientific calculators are most used by engineers, scientists, and stu-dents. Scientific calculators that are not programmable cost from $25 to $50, depending upon the number of functions. Programmable scientific calcula-tors may be as high as several hundred dollars.

Pocket computers are another step up from programmable scientific calculators. These types of calculators are actually small computers. They contain larger amounts of memory than programmable scientific calcula-tors, and the programs may actually be stored in a *programming language*, such as BASIC. Readout is still by LCD (liquid crystal display). Input is usually by a mini-keyboard - you may have to press combinations of keys to input text, rather than just one key as on the keyboard of a larger computer. Pocket computers cost anywhere from $50 to hundreds of dollars.

Electronic Keyboards

The musical world is rapidly changing to *electronic music*. Often musical scores for films are created by one man with a *synthesizer* instead of a whole studio full of musicians. Many piano manufacturers are replacing *acoustical pianos* with *electronic pianos* that create not only perfect piano sounds, but the sound of any other instrument as well.

Electronic Keyboards

The most popular consumer musical product is an *electronic keyboard* (see Figure 12-5). This resembles a piano keyboard, but has switches, buttons, and knobs above the keys. In place of the standard 88 keys of a full-sized piano, you'll find from 32 to 88 keys. Popular electronic keyboards often have 49 full-sized keys or *mini* keys, about 2/3 the width of full-sized keys.

You can play these keys to create piano sounds, if you wish, but you can also press a button and create the sound of a violin, guitar, trumpet, or mandolin. Not only can you create these sounds, but you can also press a button and hear sounds not usually associated with musical instruments - barking dogs, human voices, ocean surf, and ringing telephone bells. Some electronic keyboards offer as many as 100 separate sounds, all built into the electronics circuitry.

Also available on some keyboards are *sound sticks* which can be played on any surface (or in the air) to create percussion sounds from snare drums to bass drums. Some keyboards include the drum sounds as one of the sounds available when a key is pressed. Other keyboards have *drum pads* mounted on the keyboard, which you can play with drumsticks or your fingers. Tapping one of the drum pads will create the one or more drum types.

Another feature available on keyboards is a *rhythm section*. The rhythm section creates drum accompaniment sounds for such rhythms as rock, bossa nova, samba, waltz and others. It's like hiring a drummer to play along as you play the keyboard.

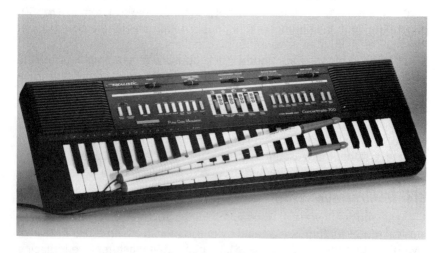

Figure 12-5.
Electronic Keyboard
Courtesy of Radio Shack
A Division of Tandy Corporation

Basic Keyboard Buying Guide

One of your first considerations in buying a keyboard is the number of sounds you want to have available. If you'll never use barking dogs, you may opt not to go for the 100 sound keyboard. Most current keyboards play from eight to 100 *preset sounds*. Fortunately, the most common sounds are the ones always included - common musical instruments and sometimes human voices.

Some electronic keyboards also *sample* sounds. They have a built-in microphone. By pressing a *sample* switch, you can capture any sound - a bark, a cough, a baby - of a few seconds duration. The amazing thing about these instruments is that the sound can then be played back at different pitches by pressing different keys on the piano! Imagine a baby's gurgle made into a Happy Birthday song.

The next thing to look for is the number of keys that can be pressed simultaneously - the number of *voices*. Unlike a normal acoustic piano, only a limited number of keys can make a sound - usually anywhere from four to eight. If you're not sure of this, simply press four to eight keys at the same time. You'll see how many keys make different sounds.

All keyboards have adjustable volume, tempo for any rhythm section, a pitch control for "tuning" the instrument, and a built-in speaker. Most have a *LINE OUT* jack for playing the instrument through an external amplifier, a headphone jack for private playing, and battery power.

Some instruments also have a *memory* that lets you record a song that you've played. Memory is usually specified in the number of notes that can be recorded. Not only can notes be remembered, but the piece can be played back at a different tempo with added rhythm.

Electronic keyboards range in price from about $50 to $500. At the top of the range are 88 key electronic pianos that deemphasize the gadgetry and stress quality piano sounds instead. About one-half of the cost of these instruments go into the "feel" of the keyboard. There's a real decision to be made here on whether you like gadgetry or old-line pianos. You'll get two different answers from innovative rock musicians and Mrs. Tucker, the nice piano teacher down the street.

MIDI Instruments

You'll find some electronic keyboards that have two or three jacks on the back labelled *MIDI OUT*, *MIDI IN*, and *MIDI THRU*. These jacks are for *Musical Instrument Digital Interface* connection, or *MIDI*. MIDI allows communication between a keyboard and electronic drums, other MIDI instruments, or even spotlights. MIDI is widely used among pop music performers. Pop musicians used to jump around on the stage plugging in *patch cords* to get different sounds. Now, they simply press buttons on their electronic keyboard.

MIDI, though, is not just used by rock stars. Using MIDI connections, you can record your electronic keyboard playing in precise detail on a *sequencer*, a sort of electronic tape recorder/computer. The piece can then be played back over the same instrument or over a different instrument, with the same or different musical sounds.

MIDI can also be driven by small computers, such as Tandy's MS-DOS systems, provided the proper *interface* is purchased. The small computer can capture all sounds played on the keyboard, save them in memory, and play them back at any later time. You can also *edit* and rearrange the songs in many different ways. You may also superimpose (dub, in effect) many different voices on top of each other, given the right *software*, or computer programs.

You don't have to use MIDI connections to enjoy an electronic keyboard, but they are something to consider when buying the keyboard. MIDI equipped electronic keyboards start at about $300.

Synthesizers

The premier electronic musical instrument is the *synthesizer* (see Figure 12-6). This is an electronic piano with additional capabilities. An electronic piano plays *preset* sounds, created by the piano manufacturer to resemble musical instruments, surf, or barking dogs. The synthesizer, however, allows the user to create his own sounds by altering electrical parameters such

Figure 12-6.
Music Synthesizer
Courtesy of Kawai

as pitch, harmonic content, sound envelope (how fast a sound rises and falls and is sustained), and others. A synthesizer can create virtually any sound - not only piano and violins, but also trains and gunshots. Sampling synthesizers are also available.

Once created, the sounds can be saved on *plug-in* cartridges, or transferred across a MIDI interface to a small computer system.

Synthesizers can play from four to 16 voices, and can even *mix* the sounds. For example, you might have four notes of a harpsichord playing at the same time that four notes of a violin are playing. With the MIDI interface, you can chain together several synthesizers to create what sounds like a small orchestra!

Needless to say, synthesizers are more expensive than their electronic keyboard counterparts. New synthesizers that include a keyboard start at about $800, with many in the $1500 range. Sampling synthesizers start at about $1800. In spite of their prices, however, synthesizers are very popular. One model, the Yamaha DX-7, has sold over 100,000 units.

Electronic Toys

There's a wide range of electronic toys available today, ranging from programmable robots to pocket games.

Robots range in price from $40 to $2000. Here we're talking about the toy variety, and not the type that plugs in circuits at a Silicon Valley (or Tokyo) manufacturing facility. Industrial robots often are of the fixed location, movable arm type. They have been preprogrammed to move through space, grasp and release objects, spin screwdrivers, position paint nozzles, and the like.

Toy robots, on the other hand, can also do some interesting things, for much less money than their industrial big brothers. Inexpensive toy robots can be moved via radio control. A small hand-held radio controller allows you to move the robot backwards, forwards, or to turn to the right or left. Some robots have built-in programming, which allows you to enter a series of commands to make the robot move forward two feet, then turn to the right, then move forward another foot, and so forth (see Figure 12-7).

Robots can also speak, uttering preprogrammed phrases such as "hello", or "what's your name?". Some robots have pincers-type hands, which can be programmed to grasp objects.

Toy robots such as the ones described cost from $50 to $500. Don't expect too much from robots such as these. If you're buying them to help with the dishes or mow the yard, you'll be disappointed. About the best such robots can do is to repeat a very rough pattern of travel, be steered by remote control, say a few preprogrammed phrases, and manage to pick up an object or two. For toys, however, they're great!

Even more elaborate robots such as Heath's Hero I and Hero 2000, costing about $1200 and $2000, respectively, are not nearly as capable as

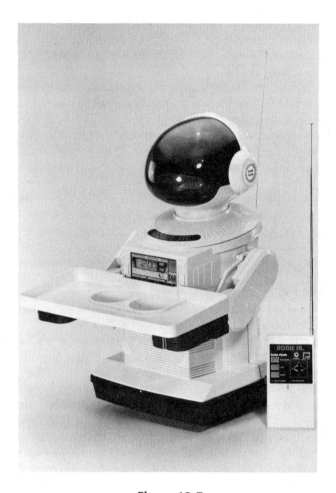

Figure 12-7.
Typical Toy Robot
Courtesy of Radio Shack
A Division of Tandy Corporation

Robbie the Robot from the movie Forbidden Planet. Although the movements may be more precise than those of toy robots, the idea is essentially
the same - a fairly dumb machine that can be programmed to perform a series
of movements. The Heath robots do have sonic distance detectors and light
detectors, but their intelligence is largely programmed by the individual who
purchases them. Programming these robots, by the way, may be quite a bit
more tedious than the toy robots above, as a type of *programming language*
has to be learned.

Toy robots in the $50 to $300 range are great for kids. If you're seriously
interested in robotics as a course of study, however, the Heath robots

provide a good training vehicle, complete with textbook and robotics course.

A variety of electronic games are available. Since most of these games have an inexpensive built-in microprocessor, the games themselves may offer sophisticated displays and decision making. Electronic chess sets can be purchased for $40 to $100 that provide a level of chess comparable to a master's level. Some have a memory capability to store the results of unfinished games!

Electronics Kits

Electronics kits provide a great learning experience for all types of electronics projects. Radio Shack kits generally provide easy-to-make connections, such as coil-spring assembly. A wire is simply inserted into a spring to make a connection as shown in Figure 12-8. This eliminates soldered connections, and makes the kits more usable by small children. All such kits are powered by low-voltage batteries, another safety factor.

Rather than having a kit devoted to a single project, these kits are more of the "30 in one" or "100 in one" variety - they contain a large number of

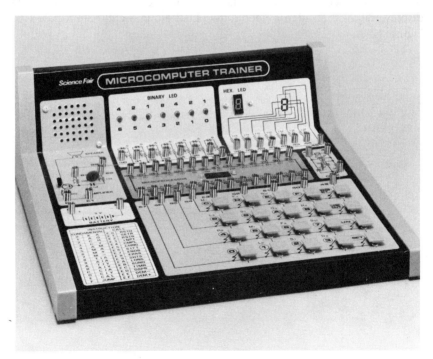

Figure 12-8.
Electronic Kit
Courtesy of Radio Shack
A Division of Tandy Corporation

generic parts which can be put together to make 30 or more projects. Good documentation is also provided for assembly and a tutorial as well.

Kits are available for electronics, mechanics, solar energy, communications, radio, and many others. Prices range from $13 to $50.

Electronics Test Equipment

Certain types of electrical test equipment are handy to have either for electrical problems around the house, or for electronics experimentation.

A *multitester* or *multimeter* is a device about the size of a book, battery-powered, with two sets of test leads, each lead ending in a needle-like probe. A multitester is used to measure voltages, current, or resistance. For example, you could easily measure your home ac voltage to see if it was within the normal limits of 110 to 120 volts with such a device. Another example of use is measuring a control voltage applied to a 24-volt ac watering sprinkler valve. A common application is to measure the *continuity* of a wire to see that it is not *open* (broken). Car battery voltage level, an indication of battery quality, can also be measured with a multimeter.

Good multitesters range from about $11 (!) to $100. The more expensive multimeters are of the *digital type* which read out directly in an LCD display, rather than with a needle deflected along a calibrated scale (see Figure 12-9).

Figure 12-9.
Multitester

Courtesy of Radio Shack
A Division of Tandy Corporation

Chapter 13
Wiring and Batteries

Puzzled by volts, amps, and resistance? In this final chapter we'll look at basic electrical concepts. You don't really need this knowledge to connect high-technology equipment, but a little knowledge of concepts certainly won't hurt, and may answer some questions for you. We'll also look at batteries to see what the best choices and bargains are.

Basic Electrical Safety Rules

Observe these safety rules while dealing with electrical appliances of any type:

- Don't handle electrical equipment while standing in water. Water and electricity don't mix.

- Never open up the back of equipment so that the wiring inside is exposed. There are extremely high voltages present in such equipment as televisions.

- Be very careful when installing antennas and cabling. Be careful not to come into contact with high-tension wires with any part of your body, aluminum ladders, antenna masts, and so forth. High-tension wires are even more dangerous than household voltages and can kill even more easily.

- When performing household wiring projects, always wear non-conductive shoes, such as rubberized tennis shoes or the like. Always turn off the circuit breaker for the wiring on which you are working - never work on a "live" circuit. Follow the "one hand in pocket, never get a shocket" rule - never complete a circuit from one hand to another hand that is in contact with a ground such as water piping or a wet floor.

Now that we've given you some basic rules, don't be too intimidated either. Most electrical equipment has been tested and designed to be safe with normal use. Just by the number of devices that are used in homes without incident, most equipment is inherently reliable. Also, almost all cabling and connections described in this book carry safe low voltages with no inherent shock hazard.

Types of Electrical Energy

There are two main types of electrical energy you'll be dealing with, so-called ac (alternating current) and dc (direct current).

Ac is generally 110 to 120 *volts*. A volt is a measure of electrical *force*. Like 110 pounds of water pressure, 110 volts exerts a great deal more force than 24 volts (24 pounds of water pressure). The voltage here "pushes" electrical *current* through a wire - the greater the force, or voltage, the more current that will be pushed through the wire.

The voltage that is present from a wall plug is 110 to 120 volts ac. It powers most household appliances and equipment, the exception being things like laundry dryers and ovens, which require 220 volts ac in many cases. All equipment discussed in this book is powered by 110 volts ac.

Much equipment *steps down* the 110 volts available from a wall socket to a lower voltage such as 24 volts ac or 12 volts dc. Lower voltage is more safe than the 110 volt wall supply, but can still cause sparks and heating effects.

Ac is used on power lines because it can be stepped up or stepped down by the use of transformers. The voltage used on high-tension lines, for example, is typically 12,000 volts. This voltage is stepped down to 440, 220, and 110 volts by transformers for commercial and household use.

Within electronic equipment, the 110 volts from a wall socket is often changed to lower voltages and then converted to *direct current*. Direct current flows in one direction only, while alternating current alternates, flowing in one direction and then in the other direction. The rate of change of alternating current is a *frequency* of 60 times per second (60 Hertz) in the United States. Other countries have different mains voltages (typically 220 volts) and different frequencies (often 50 Hertz). This means that much equipment usable in the United States will not work in foreign countries.

Electrical Current

Voltage pushes current through a wire in much the same way that water pressure pushes water through a pipe. Electrical current is made up of sub-atomic particles called electrons which travel at great speeds (close to the speed of light - 186,000 miles per second) through the wire. Billions and billions of electrons per second make up ordinary currents found in televisions and stereos. Rather than referring to a massive number of electrons as a measure of current, however, the term *ampere* or *amp* is used. One ampere represents about 16 billion *billion* electrons flowing past a given point per second.

Typical currents found in televisions and other equipment discussed in this book are from dozens of milliamperes (a milliampere is 1/1000 amp) to several amperes.

Electrical Resistance

Just as a water pipe offers resistance to the flow of water, electric conductors have resistance as well. The amount of resistance offered is affected by several things. Certain materials are better conductors and offer less resistance to the flow of electrical current than others. Gold and silver, for example, are excellent conductors with low resistance. Copper is also very good. Materials such as wood are much worse conductors than metals. Certain other materials, such as glass and plastic, are very poor conductors, called *insulators*.

The resistance of a material is measured in *ohms*. The greater the resistance of a material, the less current that will flow through it. The resistance of 1000 feet of 22 gauge copper wire is 16 ohms. Devices called resistors (available in Radio Shack) provide premeasured resistances for electronic circuits from a fraction of an ohm to millions of ohms (megohms). Sometimes it is necessary to impede the flow of current!

Ohm's Law

Learning one simple formula will provide many answers to common electrical problems. It is called *Ohm's Law*, after an early electrical scientist. It states that the current in a circuit is equal to the voltage divided by the resistance:

current (amperes) = voltage (volts) / resistance (ohms)

Using Ohm's law, it is possible to calculate many simple electrical problems. For example, suppose that you have a 6-volt battery and a flashlight lamp that the manufacturer states requires (draws) 0.1 amperes. It's resistance would be:

resistance = voltage / current = 6 / 0.1 = 60 ohms

Here the basic formula has been changed slightly to find resistance given voltage and current. Another formula from the basic formula is

voltage = resistance x current

Power

Power is the rate of doing work. It's expressed in *watts*. In electrical terms, it's current times voltage. One ampere following through a circuit with a voltage of one volt produces one watt of power.

power (watts) = voltage (volts) x current (amperes)

The same flashlight bulb above will use power of

power = 6 x 0.1 = 0.6 watts

Basic Home Wiring

Most electrical wall outlets provide 110 volts ac and from 10 to 15 amperes of current. This means that

power = 110 x 10 = 1110 watts

or

power = 110 x 15 = 1650 watts

House wiring is divided into many circuits. If you have a fuse box, you'll find a fuse for each circuit. If you have circuit breakers, there'll be a circuit breaker for each circuit. Usually, circuits are divided logically on the basis of how they will be used and how much power is required. Kitchen stoves and ranges, for example, are on a separate circuit from a bedroom.

The maximum current permitted on a circuit is done for a reason. If the current exceeds the maximum, the current in the wire will cause heating effects (power). Therefore, the fuse or circuit breaker protects excessive current (high powered appliances or equipment) from being used.

Most audio, video, and telephone equipment described in this book is relatively low power. Typical televisions, for example, are about 150 watts, and require

current = power / voltage = 150/110 = 1.4 amperes

Even audio amplifiers that provide 100 watts per channel are low-power devices. The stereo audio power is not continuous and the *average* power is at a much lower level than 100 watts.

However, you should be aware of the maximum current provided by a house circuit and connect equipment accordingly. A 10-ampere circuit with a television, video cassette recorder, audio amplifier, and cassette deck may overload a circuit already connected to 600 watts of lighting (5.5 amperes) and 300 watts of computer equipment (2.7 amperes).

Most newer ac plugs are *polarized*. This means that the plug is *keyed* so that the prongs of an appliance can go in only one direction. One side of the ac plug should be *grounded*. A ground is a common circuit, often connected to earth ground. The other lead of the circuit is "hot" or ungrounded.

Grounding is an important safety feature. If a piece of electrical equipment should short, the hot side of the circuit (110 volts ungrounded) may contact the metal conducting case of the equipment. If you come in contact with the case and also happen to complete the circuit to ground via water pipes or a wet concrete floor, you might receive a shock. With grounded equipment, however, this danger is minimized. The circuit breaker or fuse will blow as excessive current flows from the hot side to ground.

Grounding is used for another good reason, however. Grounded devices have less electrical noise. This applies to audio, video, and telephone equipment - any device where there is an audio or visual output. Make certain that devices are grounded properly to minimize noise.

A small test device called a *circuit tester* is often sold at hardware stores. This device has a readout of green and red LED (light emitting diode) indicator lamps. It can be plugged into each wall socket outlet to tell you whether the outlet is properly grounded and connected.

To extend ac power, use polarized or grounded (3 outlet) extension cords. Often it's a good idea to use *power strips*. These are long boxes with multiple outlets. Many times the box has a switch with a power light on it and may also have a built-in circuit breaker. The advantage of a power strip is that the plugs for an audio entertainment center or video equipment can all be connected to the one source and then switched on or off as a single unit, instead of having to turn individual switches on and off.

Running Cables and Lines

It's often necessary to run cables and lines over dozens of feet to connect security systems, audio speaker cables, and the like. There are a few ground rules to follow:

- Always stay away from live ac power circuits. Cables and lines should be physically separated from ac circuits.

- It's always proper to use larger diameter wire than required, but do not use smaller diameter wire than that recommended. A 22 gauge "hookup" wire is about the smallest diameter recommended for general use. Use *stranded* copper wire, if the connections will have a lot of kinks and bends; stranded wire breaks much less easily.

- For audio and video applications, always use shielded cable of the proper type (audio or video).

Batteries

Batteries are used in all types of audio and video equipment, even that which is powered primarily by ac power. With *intelligent* devices, batteries are required to maintain the "programming" of the device after main power is turned off or disconnected. Remote devices - such as infrared control devices - also need battery power.

Batteries are classified according to size, shape, and type of material used to generate the energy.

Battery Sizes and Shapes

Figure 13-1 shows the actual sizes of common battery types. Size AAA and AA are often used in audio or video equipment where not too much current is required. They provide voltages of about 1.25 volts each. C cells and D cells store more energy and are typically used in flashlights. Again, their voltage is 1.25 volts per cell. N sized cells are often used in cameras or photographic equipment. These batteries are cylindrical shaped and again provide about 1.25 volts. The 9 volt "transistor" battery is rectangular and is often used in small transistor radios. "Lantern" batteries are used in larger flashlights or lanterns or in other types of lighting equipment and provide voltages of 6 volts.

Battery Capacity

Battery energy requirements are listed in volts and *ampere-hours*. Ampere-hours as the term implies, are a measure of how much current can be supplied over a certain time. A battery with a 1.0 ampere-hour rating, for example, can supply 1/2 ampere for two hours, or 1/4 ampere for four hours.

Types of Battery Material

There are generally four types of batteries commonly available, general-purpose batteries, heavy-duty batteries, alkaline batteries, and nickel cadmium batteries. Their prices increase in that order.

General-purpose batteries are usually carbon zinc batteries. This type works best with low or moderate power requirements. Carbon zinc batteries don't withstand heat or cold very well and do not have a long shelf life. On the other hand they are inexpensive.

Heavy-duty batteries are related to carbon zinc batteries but use zinc chloride as the material to generate energy. Zinc chloride batteries cost about 60% more than carbon zinc batteries. They perform better at low temperatures (high temperatures are still a problem) and provide more energy over about 50% more time than carbon zinc batteries.

Alkaline batteries cost about three times as much as the carbon zinc general-purpose batteries, but provide plenty of current for high-current equipment, such as motorized toys and cassette recorders. Alkaline batteries may provide seven or eight times as much life as the general purpose type. Their storage life is very good, and they can withstand both low and high temperature extremes. In general, alkaline batteries are much more cost efficient than the other types in spite of their higher initial cost.

Rechargeable Batteries

The three battery types above are "throw-away" batteries. After their voltage falls, they become unusable and must be thrown away. Nickel cadmium or NICAD batteries, however, are rechargeable and reusable almost indefinitely. (NICADs are the same physical size as other types of batteries.) The cost of nickel cadmium batteries is high - they are about twice that of alkaline batteries (six times that of general-purpose batteries). However, they can be charged overnight and reused many times. Nickel cadmium batteries typically do not last as long as alkaline batteries, and their voltage may be somewhat lower.

High-capacity NICADs provide a longer life than normal NICADs at a higher cost. These NICADs may run more than twice as much as the normal NICADs.

Battery chargers for NICAD batteries cost about $10 to $30, with the average cost being on the order of $13. These battery chargers run off of ac power, but use a minimal amount of current - only a few pennies - for

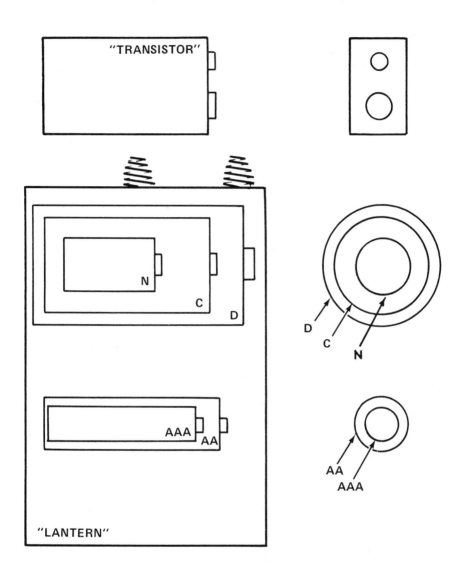

Figure 13-1.
Battery Sizes

recharging. Often the rechargers can recharge several sets and sizes of batteries at once. Assume that the battery charger costs $13 and two "D" type NICADs cost $3. The $16 cost of the batteries and recharger is recovered after about 13 recharging cycles if alkaline batteries are replaced by the NICADs. With 10 NICADs of various types costing an average of $2.50 each, the $38 initial investment in batteries and recharger is recovered in about four recharging cycles, assuming that alkalines are again replaced.

Of course, the other side of the story is that recharging is somewhat of a nuisance. It's much easier just to throw away a set of batteries and pop a new set in.

Index

I

Impedance, of speaker system:
18-19
Interlace, television: **53-54**
Intermodulation distortion, in
stereo system: **11**

K

Key switches, for alarm
system: **99-100**
Keyboard, electronic,
buying guide: **149**
general: **147-151**
Kits, electronic: **153-154**

L

L-pads, for speakers: **42**
Light beam devices: **103**
Loudness control, in stereo
receiver: **13**
Loudness effect, in stereo
system: **13**
Luminance, of televisions: **59**

M

Matching transformer, for
antenna: **44, 46**
Microphones, for tape recording:
31
MIDI keyboards: **149-150**
Mixing equipment: **36-37**
Modular plugs, telephone:
114-115
Monitors, television,
audio connections: **48-49, 85**
general: **56**
Motion detectors, for alarm
system: **100, 102**
MTS Stereo,
converters: **91**
general: **57-58**
VCRs: **70**
Multimeter: **154, 155**

Multiple-line phones: **109**
Multitester: **154, 155**

N

NICAD batteries: **162, 164**

Noise, in audio system, getting
rid of: **50-51**
Normally-open, normally-closed
circuits, for alarm: **96-97**

O

Ohm's Law: **159**
On-screen programming, in
VCRs: **69**

P

Paging devices: **124-128**
Phasing of speakers: **42**
Phone dialers: **103-104**
Phone plugs: **79-80**
Phono plugs/jacks: **77-78, 79**
Picture quality,
televisions: **58-59**
VCRs: **68**
Picture-in-picture,
television: **59**
VCRs: **68**
Placement, of speaker system:
19-20
Plug 'N Power devices: **104-106**
Pocket pagers: **124-126**
Power required, for stereo
system: **8-9**
Power, audio,
speaker system: **18**
stereo system: **8**
Power, electrical, definition: **159**
Programming, in VCRs: **69**
Pulse vs. tone dialing: **108**

R

Radar detectors: **134-136**